Advanced Multiphasing Switched-Capacitor DC-DC Converters

Nicolas Butzen • Michiel Steyaert

Advanced Multiphasing Switched-Capacitor DC-DC Converters

Pushing the Limits of Fully Integrated Power Management

 Springer

Nicolas Butzen
Intel Labs
Intel Corporation
Portland, OR, USA

Michiel Steyaert
ESAT-MICAS
KU Leuven
Leuven, Belgium

ISBN 978-3-030-38737-2   ISBN 978-3-030-38735-8   (eBook)
https://doi.org/10.1007/978-3-030-38735-8

This Springer imprint is published by the registered company Springer Nature Switzerland AG.
The registered company address is: Gewerbestrasse 11, 6330 Cham, Switzerland

Preface

Power converters are an essential part of a global infrastructure upon which much of today's society depends. For DC–DC converters used in electronics, there has been a strong drive towards integration on microchips, ideally together with their load. In addition to benefits in terms of size and cost, the resulting fully integrated power converters are key in reducing losses related to the power delivery network and in enabling granular dynamic voltage scaling in digital circuits, both of which would lead to substantial improvements in system efficiency. Furthermore, because many of today's central processing units and systems-on-a-chip are limited by their power-density, this efficiency gain could directly be leveraged for performance.

Switched-capacitor DC–DC converters have emerged as a promising candidate for integration because both switches and capacitors are readily available in today's integrated technologies. That being said, even for these converters the monolithic context has proven a challenge. Partly due to the large **parasitic coupling** and the limited **capacitance density**, the efficiency and power-density that can be obtained are constrained as well. In addition, this type of converter inherently has a very narrow **conversion ratio range** close to a fixed rational conversion ratio. Naturally, this is problematic for battery-connected devices or dynamic voltage scaling. The main aim of this work is to alleviate these limitations through the concept of advanced multiphasing. Here, multiple converter cores are put in parallel and actively interact with each other over several phases to improve their capabilities. As such this work can also be considered an exploration into the so-far largely unstudied phase domain.

This book gives an overview of the fundamental concepts behind monolithic switched-capacitor converters and points out that, after optimization, there is distinct low and high power-density regime at which they can operate. In the former, the converter's efficiency reaches a maximum, while in the latter an efficiency-power-density trade-off can be witnessed. Three figures of merit are introduced that compare a converter's performance to these theoretical regimes or evaluate a converter to reduce power delivery network-related losses. The theory is expanded with the presentation of a voltage-domain analysis that leads to a fundamental law of conventional switched-capacitor converters. Relating several flying capacitor attributes to the conversion ratio, this law indirectly demonstrates the key advantage multiphase converters have over two-phase converters.

With the goal of reducing **parasitic coupling** losses, a first advanced multi-phasing technique, called scalable parasitic charge redistribution, is proposed where the charge on the parasitic coupling is continuously recycled between out-of-phase converter cores through several charge redistribution buses. Because this technique removes the previously established efficiency limit, the basic loss model is updated with transistor leakage to establish a new fundamental maximum. Thanks to the technique, a converter is realized that reduces its parasitic coupling losses tenfold and achieves a record 94.6% efficiency. Also, because the charge redistribution buses have a DC voltage, they are investigated as a means to supply power to circuitry within the converter itself.

Stage outphasing and multiphase soft-charging are two more techniques that instead focus on the limited **capacitance density** by spreading charge transfers between capacitors out over multiple smaller and more efficient steps. The result is that the effective capacitance density of the converter is improved, which is useful for high and low power-densities. Both techniques are shown to work for several topologies, but are especially beneficial in combination with Dickson converters with larger conversion ratios. An implementation in a baseline integrated technology verifies the working principle of the techniques by obtaining 60% larger effective capacitance density and a record 82%-efficiency 1.1 W/mm^2-power-density combination.

Conventional switched-capacitor topologies minimize the voltage swing on their capacitors to reduce charge-sharing losses and arrive at an efficient operation, though only for a narrow conversion ratio range. In this book, a fundamentally new type of switched-capacitor converter is introduced with large voltage swing capacitors that is made efficient using soft-charging enabled by advanced multiphasing. A particular topology is found to behave like a gyrator with a **continuously scalable conversion ratio**, which is the first time a purely capacitive DC–DC topology achieves this feat. A realized converter implements this topology and obtains the largest efficient conversion ratio range in the literature.

Portland, OR, USA   Nicolas Butzen
Leuven, Belgium   Michiel Steyaert

Contents

Abbreviations

AC	Alternating current
AM	Advanced multiphasing
BEOL	Back end of line
BOM	Bill of materials
BP	Bottom plate
CMOS	Complementary metal-oxide-semiconductor
CPU	Central processing unit
CRB	Charge redistribution buses
CRS	Charge redistribution steps
CRT	Charge redistribution transistors
DC	Direct current
DNW	Deep N-well
DVFS	Dynamic voltage and frequency scaling
DVS	Dynamic voltage scaling
EEF	Efficiency enhancement factor
EMI	Electromagnetic interference
ENIAC	Electronic numerical integrator and computer
FF	Flip-flop
FoM	Figure of merit
FSL	Fast-switching limit
GPIB	General purpose interface bus
IC	Integrated circuit
IoT	Internet of Things
iVCR	Ideal voltage conversion ratio
I/O	Input/output
KCL	Kirchoff's current law
LDO	Low-dropout
MIM	Metal-insulator-metal
MOM	Metal-oxide-metal
MOS	Metal-oxide-semiconductor
MOSFET	Metal-oxide-semiconductor field effect transistor

MSC	Multiphase soft-charging
NMOS	N-channel metal-oxide-semiconductor
NOC	Non-overlapping clock
PCB	Printer circuit board
PDN	Power delivery network
PMOS	P-channel metal-oxide-semiconductor
PMU	Power management unit
PoL	Point-of-load
PWM	Pulse-width modulation
SAR	Successive approximation register
SC	Switched-capacitor
SiP	System-in-package
SMD	Surface-mount device
SO	Stage outphasing
SoC	System-on-a-chip
SOI	Silicon-on-insulator
SPCR	Scalable parasitic charge redistribution
SR	Set-reset
SSL	Slow-switching limit
TDP	Thermal design power
TP	Top plate
VCR	Voltage conversion ratio
VDA	Voltage-domain analysis

List of Symbols

α_C	Capacitance density per unit area
α_{CE}	Capacitor energy density per unit area
α_{Cpar}	Parasitic coupling capacitance density per unit area
α_l	Leak-to-conductance conductance ratio
$\alpha_{l,c}, \alpha_{l,nc}$	Leak-to-conductance conductance ratio in (non-)conducting state
α_{par}	Ratio of the parasitic coupling and the capacitor's capacitance
α_S	Switch GV^2 density per unit area
$\delta(v)$	Dirac delta distribution
$\Delta v_i, \Delta v_{ij}$	Normalized voltage swing of parasitic capacitor i (phase j)
ΔV	Full voltage swing of a capacitor
ΔV_{ij}	Voltage swing of parasitic capacitor i phase j
ΔV_{pi}	Voltage drop at output port i
ϵ	Permittivity
η	Efficiency
ϕ_i	Converter phase i
$a_{c,i}, a_{c,ij}$	Charge multiplier element of capacitor i (phase j)
$a_{R,i}, a_{R,ij}$	Charge multiplier element of switch i (phase j)
A_{conv}	Converter die area
A_{Ctot}	Total capacitor die area
$A_{C\%i}$	Capacitor die area share of capacitor i
A_{conv}	Converter die area
A_{Rtot}	Total switch die area
$A_{R\%i}$	Switch area share of switch i
b_i	Number of activations per cycle of switch i
B_i, Bi	CRBs or bottom-side intermediate soft-charging nodes
C_{chan}	Channel to bulk capacitance
C_{DC}	Output capacitance
C_{fly}	Flying capacitor
$C_i, C_{\%i}$	Capacitance (share) of ith capacitor of converter
C_{nwpsub}	N-well to p-substrate capacitance
C_{ox}	Transistor oxide capacitance

$C_{par,i}$	Parasitic capacitance of ith capacitor of converter
C_{pwdnw}	P-well to deep N-well capacitor
C_{tot}	Total converter capacitance
CRS	Number of charge redistribution steps in SPCR
$D_{cond,i}$	Total conduction duty cycle of switch i
D_j	Duty cycle of converter phase j
E_{cs}	Energy lost by charge-sharing
E_F	Electric field strength
E_{int}	Intrinsic energy lost per cycle
$E_{on,i}$	Activation energy of switch i
E_{on}/A_{on}	Transistor activation energy per unit die area
$E_{on}R_{on}$	Transistor activation energy-resistance product
E_{supply}	Energy taken from a supply
f_{clk}, F_{clk}	Converter control or external reference frequency
f_{sw}	Switch frequency of converter
$F_i, F_i^{(j)}$	Fibonacci number (order j)
$G_i, G_{\%i}$	Conductance (share) of ith switch of converter
G_{tot}	Total converter switch conductance
G'_{tot}	Total converter switch conductance divided by converter area
H_{conv}, H_{cap}	Cumulative power distribution area of converter/flying capacitor
$iVCR$	Ideal voltage conversion ratio
I_C, I_{Cj}	Time-averaged current through a capacitor in a phase
I_{Cij}	Time-averaged capacitor current through capacitor i phase j
I_{gate}	Transistor gate tunneling current in conducting state
I_{load}	Load current
I_{out}	Output current of converter
I_{pi}	Current drawn from port i
I_{rev}	Transistor gate tunneling current in non-conducting state
$I_{R,ij}$	Current through switch i in phase j
I_{subt}	Channel leak current in transistor
I_{supply}	Supply current
$I(v)$	Current distribution in voltage domain
$I^*(v)$	Cumulative current distribution in voltage domain
K_C, K'_C	SSL topology factor
K_{par}, K'_{par}	Parasitic coupling topology factor
K_S, K'_S	FSL topology factor
K_{trans}, K'_{trans}	Transistor driving topology factor
P_D	Converter power-density
P_{FSL}	FSL conduction power loss
P_{in}	Input power
P_{int}	Normalized intrinsic converter loss
P_{leak}	Total leakage power loss
$P_{l,c}, P_{l,nc}$	Leakage power loss in (non-)conducting state
P_{loss}	Total converter power loss
$P_{loss,ij}$	Power loss of switch i in phase j

P_{node}	Output power at a node
P_N	Normalized total converter power loss
P_{out}	Output power of converter
P_{par}	Parasitic coupling power loss
P_{PDN}	Normalized power loss related to the power delivery network
P_{SSL}	SSL conduction power loss
P_{trans}	Transistor-driving power loss
Pi	Converter phase i
$P^*(v)$	Cumulative power distribution in voltage domain
$q_{c,i}, q_{c,ij}$	Charge transferred by capacitor i (phase j)
$q_{out}, q_{out,i}$	Output charge per cycle (at output i)
$q_{R,ij}$	Charge passing through switch i in phase j
R_{bias}	Bias resistance
R_{FSL}	Fast-switching limit output resistance of converter
R_{out}	Output resistance
R_{SSL}	Slow-switching limit output resistance of converter
S	Number of soft-charging phases
S_1, S_2, S_3, S_4	Switches
t	Time
T_i, Ti	Top-side intermediate soft-charging nodes
v_{Ci}	Normalized capacitor voltage
v_{bij}	Normalized bottom-plate voltage of capacitor i in phase j
v_{Ri}	Normalized switch block voltage
V_b, V_{bj}	Bottom-plate voltage of capacitor (phase j)
V_{bias}	Bias voltage of DNW in MOS capacitor implementation
$V_{B,i}$	Voltage of CRB
V_C, V_{Ci}	Capacitor voltage
V_{fb}	Feedback voltage in converter control loop
V_{gst}	Transistor gate-source voltage, minus the threshold voltage
V_{high}	High voltage of parasitic coupling swing
V_{in}	Input voltage of converter
V_{low}	Low voltage of parasitic coupling swing
V_{max}	Device maximum voltage rating
V_{out}	Output voltage of converter
V_{pi}	Voltage at output port i
V_{ripple}	Ripple voltage at converter output
V_{Ri}	Switch block voltage
V_{ss}	Ground voltage of converter
V_{tj}	Top-plate voltage of capacitor in phase j
$V_{trans,i}$	Transistor driving voltage
V_T	Transistor threshold voltage
Y	Gyrator conductance
z_{ij}	Transimpedance element
$\mathbf{Z}$	Impedance matrix of a MIMO converter
Z_{bias}	Bias impedance

Chapter 1
Introduction

Since the invention of the solid-state transistor, and the subsequent digital revolution, society has changed dramatically. For example, less than two decades after Tim Berners-Lee and Robert Cailliau made the very first website available to the general public [BLC90], approximately two thirds of people worldwide now use the internet at least occasionally to check their e-mails, look up things online, or access social networks [Pew16]. Modern economies, especially in developed countries, are in similar fashion shifting rapidly to information technologies. Under the hood of these technologies, there is an extensive physical infrastructure on which it depends. Major internet companies have thousands of data centers all over the world, which connect to each other and consumers through global communication networks. Cheap smartphones are similarly available worldwide and act as internet access points for billions of people.

In all of these systems an essential part is played by energy. Energy itself comes in many forms. For electrical energy in particular, one must take into account the energy's potential or voltage and its frequency in case of alternating current (AC) or lack thereof for direct current (DC). In addition, energy needs to be supplied at a sufficiently high rate to meet the specified power requirements. Unsurprisingly, due to the many parameters involved, the energy's preferred form depends heavily on the context. For example, while transportation of energy is preferably done at higher voltages to minimize ohmic losses, consumption usually requires much lower voltages for safety, smaller size, and, in some cases, better performance. As a direct consequence there is a need for interfacing systems that transform electrical energy into the desired form. These systems are called power converters. From energy generation, say a power plant, to energy usage, there are many such converters in its path. Generally speaking, the closer one gets to where energy is used, the smaller and more low-power the converters are.

This work focuses on the smallest of DC to DC power converters, those that are integrated entirely on a single microchip, ideally together with its load. It is the ambition of this book to demonstrate, through both analysis and experimental

© Springer Nature Switzerland AG 2020
N. Butzen, M. Steyaert, *Advanced Multiphasing Switched-Capacitor DC-DC Converters*, https://doi.org/10.1007/978-3-030-38735-8_1

verification, how the performance of such fully integrated power converters can be improved by making better use of the time-, or phase domain.

This introductory chapter will take a closer look at the historical background of integration, and the general driving factors towards fully integrated power converters specifically. Afterwards, the challenges with respect to integrating power converters, and switched-capacitor (SC) converters in particular, are discussed. Next, the general idea of advanced multiphasing (AM) is introduced, together with its effect on the design space of monolithic SC converters. Finally, this chapter will conclude with a general outline of this book.

1.1   The Road to Integration

1.1.1   Historical Background

The Integrated Circuit

The first electrical computers, with an early programmable and Turing-complete computer being the Electronic Numerical Integrator and Computer (ENIAC) shown in Fig. 1.1, relied on vacuum tubes [GG46]. The problem with these tubes was that they were quite large and consumed a lot of power. At the same time, they were not reliable: The ENIAC reportedly needed a tube replaced once every 2 days [Ran06]. Despite their practical issues, the first generation of electrical computers did contribute greatly to multiple areas, including the United Kingdom's code breaking effort during the Second World War [Cop06], and the USA's development of thermonuclear weapons [Gol72].

Realizing this potential, several groups started researching alternatives for the vacuum tube. In 1947, Bardeen and Brattain from Bell Labs demonstrated a point-contact transistor made out of two closely placed gold contacts on a semiconductor substrate [BB48]. Their manager, Shockley conceived the bipolar transistor the next year, based on his theory of p-n junctions [Sho76], and shortly after, Dummer envisioned electronic circuits "in a solid block with no connecting wires" [Dum53]: The integrated circuit (IC) was born [Kil76].

Not only was the integrated circuit much smaller than the sum of its elements separately, it also provided a platform for future miniaturization thanks to the planar manufacturing process [Hoe60]. Since then, every few years, a new integrated technology with smaller transistors has been developed.

Moore's Law

First proposed by Gordon Moore in 1965, Moore's law states that the number of transistors on integrated circuits increases exponentially over time [Moo65]. Interestingly enough, his observation was not that it *could* be done from a technical

Fig. 1.1 The ENIAC computer being programmed

point of view—although this of course did play an important role—but rather that it would be economically sensible to do so based on the decreasing cost per transistor with every smaller transistor technology. With an exponential increase in the number of transistors, the processing power of ICs was predicted to increase exponentially as well.

This effect was further enhanced by Dennard's guidelines for transistor scaling [DGY⁺74]. He noted that ICs could run at consistently higher clock frequencies with every technology generation while keeping their power-density constant, if their voltage and current scaled with the same factor as the physical dimensions of the transistors. Thus, every next generation of ICs did not only have more transistors, the transistors themselves would also be faster than the previous generation.

For several decades, Moore's law with Dennard scaling led to steady exponential improvements in the raw computing power of digital processors [Rup18]. At some point, though, supply voltages were getting close enough to the transistors' threshold voltage that leakage currents became substantial, and thus Dennard scaling had to be abandoned to prevent leakage power from growing beyond dynamic power consumption [EBA⁺13]. Instead, supply voltages scaled at a slower rate, but this also caused ICs' power-densities to increase. With their power consumption increasing, eventually a point was reached where high-end central processing units (CPUs) generated so much heat, that they could barely be cooled enough to avoid a thermal runaway. As a result, the clock frequency of CPUs could no longer be increased naturally, and single-thread performances started to stagnate [BSW15].

While Moore's law continued by shifting focus to multi-core processors, it was clear that a CPU's performance was (and still is) limited by its power-density, or rather its thermal design power (TDP). Research consequently shifted towards making ICs more efficient. After all, any increase in system efficiency can be leveraged for better computational performance. It is in this context that integrated power management units (PMUs) started to garner a lot of attention [BSS16].

1.1.2 Key Motives for Integrated Power Conversion

In many ways, the drive towards integrated power converters or PMUs in general shares a lot of the motives with past efforts towards integrating memory, accelerators, graphic processors, modems, and so on, which led to the current standard of systems on a chip (SoCs) in mobile applications. Broadly speaking, these motives can be divided into three categories.

Size

Naturally, shifting any system block from a separate IC to the main IC reduces the required printed circuit board (PCB) area and thus the total system size. Because power converters typically require multiple additional discrete passive devices such as capacitors and inductors, the benefit is even more pronounced. Inductors also tend to have higher profiles, which might be an issue in some use cases.

A smaller system size, in turn, opens up new or better applications. Smart watches, for example, would have been impossible to realize as they are known today, if not for their high level of integration.

As more and more systems are integrated, one eventually reaches a point where the PCB is no longer required: Either the system fits entirely on a single microchip, or the few dies that remain can be bonded together with a small battery [FKC+13]. At this level, the total volume is expressed in cubic millimeters, rather than cubic centimeters. In addition, because it would be impractical to charge these systems sporadically, they need to get their energy from their surroundings through energy scavenging. As such, integrating PMUs that can properly deal with the energy harvester was a major condition for their feasibility, and remains a popular research topic to this day.

Cost

As discussed previously, a higher level of integration reduces the PCB size and the number of external components, usually referred to as the bill of materials (BOM). Both have a direct, positive impact on the system's cost of manufacturing and assembly. Moreover, by reducing the number of ICs on a PCB, the required number of connections between those ICs drops as well. The PCB might consequently not need as many layers, thus making it cheaper.

Microchips, especially those using silicon CMOS processes, are produced in massive volumes which drives down their price. Furthermore, because the cost per transistor has continued to decrease with technology scaling [Hol16], the economical incentive for higher integration has been increasing as well.

Performance

Due to the shorter distances that need to be traversed, functional blocks that are placed close to one another communicate faster while using less energy. As such, it comes as no surprise that the millimeter scale separations or less in microchips have a significant advantage compared with PCB-level designs. For fully integrated PMUs, this proximity to their load translates into better regulation and can enable efficiency improvements like fast dynamic voltage- or dynamic voltage and frequency scaling (DVS/DVFS) in digital processors [CB95].

Monolithic PMUs also hold significant promise to improve the transport of energy to integrated digital circuits by reducing so-called power delivery network (PDN) losses. To demonstrate this, consider the schematic representation of a typical PDN in Fig. 1.2a. Using an external DC–DC converter, the IC has a certain total intake current, I_{load}, that induces both static- and dynamic voltage droops in the PDN proportional to the intake current itself. To cope with these voltage

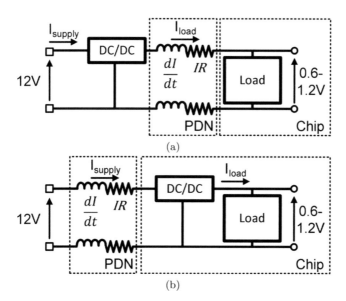

Fig. 1.2 Simplified schema of an example power delivery network (PDN) to an integrated load using (**a**) an external- and (**b**) an on-chip DC–DC converter

droops, digital circuits need to be designed with a worst-case voltage margin in mind to guarantee their correct operation [CFM⁺10]. With load voltages of digital processors continuing to decrease over time, while simultaneously increasing their total power consumption, these effects have become large loss contributors: Already in 2014, the effective efficiency of energy transport to the main SoC of smartphones was estimated to be less than 70% [Car14]. As a result, the percentage of input/output (I/O) pins of high-end processors that is dedicated to power delivery has continued to increase as well, in an attempt to limit the PDN's impedance [SMCL11, AKK⁺13]. In contrast, shifting the DC–DC converter to the same die as the processor, as shown in Fig. 1.2b, reduces the intake current together with the current-induced droops by the on-chip voltage conversion ratio (VCR) [CFM⁺10]. The voltage margins can consequently be minimized to significantly improve the full-system efficiency, which can in turn be leveraged for raw performance.

Finally, fully integrated PMUs are a key enabler for extensive granularization of voltage domains in today's SoCs and processors [SVBM⁺11, BS11b]. The central idea is simple: if every core or functional block of an IC would be supplied a voltage that is just high enough to fulfill its task in a given time, independent of the rest of the IC, then a substantial amount of energy can be saved [AKK⁺17]. Depending on the workload, a processor's efficiency could be increased by as much as 21% if per-core DVFS is used [KGWB08]. Because this does mean that every granular block requires its own voltage domain and thus voltage regulator, realizing a high level of granularization is impractical with external PMUs.

1.2 Integrated Switched-Capacitor Converters

1.2.1 Why Switched-Capacitor?

By far the most popular type of point-of-load (PoL) converter is the buck converter. Its popularity can at least in part be attributed to its remarkable simplicity. In its most common mode of operation, the buck converter generates a pulse-width modulated (PWM) signal using just two switches, which is then passed on to an inductive low-pass filter to arrive at a suitable output voltage. However, the lack of quality of integrated inductors has posed a serious problem for their full integration. To make things worse, this quality issue appears to be tied to the inductor's small size and thus fundamental in nature [SRSK16]. As a result, most inductive converter designs in the literature have opted for a system-in-package (SiP) rather than a SoC approach where the inductor is not integrated on the die itself, but made out of bondwires [WCS07], extra back end of the line (BEOL) metal and/or magnetic layers [KCB⁺15], PCB tracks [SS15b], or a surface-mount device (SMD), possibly mounted directly on the silicon [SDS17].

Switched-capacitor (SC) converters, in contrast, rely only on switches and capacitors, both of which are readily available in modern CMOS processes.

Furthermore, their quality is expected to continue to improve from one technology generation to the next, because of the key importance of transistors and capacitors in digital circuits [ITR15]. That being said, the quality of monolithic capacitors is still substantially worse than their external counterparts, and inherently limits the performance of this type of converter.

An additional advantage of SC converters is that they do not rely on a magnetic field for their operation. This has led to some researchers promoting their use in small form-factor applications where electromagnetic interference (EMI) is a concern, such as hearing aids [LVJ17]. An overview of the benefits of integrating SC PMUs is given in Fig. 1.3 for an example SoC.

Regardless of the many advantages, though, the fully integrated SC converter must satisfy a number of properties to be viable. While different applications will, of course, emphasize different attributes, efficiency and power-density—in order to minimize the converter's area overhead—are almost universally important. In addition, when the goal is to reduce PDN losses, the higher the VCR, the larger the potential system-efficiency improvement. Finally, the SC converter should be able to handle a sufficiently large VCR range to either deal with changing battery voltages or to enable techniques such as DVFS, and only require commonly available devices to enable wide-spread use.

1.2.2 Switched-Capacitor Converter Design Space

While there are multiple characteristics or specifications one can attribute to a DC–DC power converter, this section will focus on three which were identified as critical in the previous discussion: Efficiency, power-density, and VCR range.

Efficiency is usually defined as the ratio of the useful power versus the total power. What exactly constitutes useful power depends on the context. The most common definition, though, is as follows:

$$\eta = \frac{P_{out}}{P_{in}}, \tag{1.1}$$

where η is the efficiency, and P_{out} and P_{in} are the power delivered to the output and drawn from the input, respectively. Power-density, on the other hand, is a measure of the converter's output power normalized by its size:

$$P_D = \frac{P_{out}}{A_{conv}}, \tag{1.2}$$

with P_D the power-density and A_{conv} the converter's die area.

By plotting the highest obtainable efficiency after optimization for each power-density point, as shown in Fig. 1.4, the monolithic SC converter's design space is revealed. This space is bound by two asymptotic limits [LSA11]. At low power-

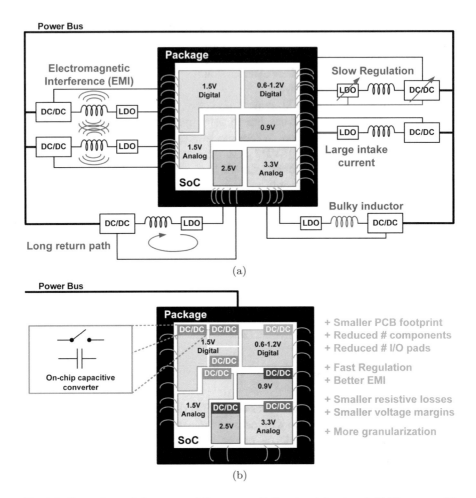

Fig. 1.3 Comparison of the power delivery to a SoC using (**a**) external PMUs versus (**b**) monolithic switched-capacitor PMUs

densities, a combination of losses related to charging the capacitor's parasitic coupling, C_{par}, and the total flying capacitance, C_{tot}, itself, forms an effective efficiency-ceiling. Transistor leakage is an additional contributing factor in this region, but typically of lesser importance [BS16a]. For increasing power-densities, eventually a point is reached where the transistor charging losses, related to $E_{on}R_{on}$ shown in Fig. 1.4, overshadow the parasitic capacitor losses. As a result, a clear trade-off between efficiency and power-density is established. The limits themselves are dependent on the technological parameters highlighted in Fig. 1.4. The performance that one can obtain is thus set by the quality of capacitors and transistors available to him.

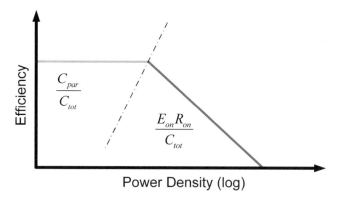

Fig. 1.4 Efficiency versus power-density design space of a fully integrated SC converter

In terms of VCR range, switched-capacitor converters are severely limited, but not by technology. A conventional SC topology can inherently only convert a single fixed rational conversion ratio. Any deviation from said conversion ratio has to be paid for with efficiency. The most common approach to deal with this restriction is to combine multiple topologies into a single converter and switch between these topologies when needed. However, this solution also has to downside of increasing the system complexity, and reducing the efficiency and power-density for the full VCR range [SRS18].

1.2.3 Advanced Multiphasing

The use and exploration of advanced multiphasing is based on three key observations [BS16c, BS17b]. First, technology scaling is driven by a reduction of cost and power per transistor. As such, complexity has become very cheap and efficient in advanced CMOS technologies. In addition, monolithic SC converters can easily be fragmented into several converter cores with little area overhead and thus without negatively impacting the efficiency and power-density of the whole. The number of components integrated on a chip is, in essence, not as important as the total area that is consumed. Finally, technology scaling has also led to modern CMOS technologies that are able to handle very high clock frequencies [ITR15]. The subsequent discrepancy between the typical two-phase converter frequency and the technology frequency means there is still room in the largely unexplored time- or phase domain [SBM+].

Some of these observations have already been exploited to some extent in earlier work. By using three-phase converters, for example, higher conversion ratios have been shown to be realizable using the same number of components [KP14]. This phase-domain space is also used in the multiphasing technique. Here, a SC converter

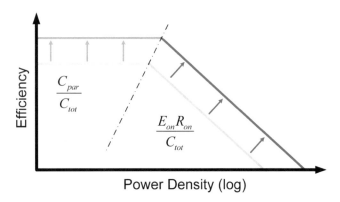

Fig. 1.5 Effect of advanced multiphasing on the efficiency versus power-density design space of a fully integrated SC converter

is split into multiple cores that switch consecutively, or out-of-phase, rather than at the same time, thus reducing the output voltage ripple [BS09, BS11b, Piq12].

Advanced multiphasing simply extends on the multiphasing concept by also making the cores interact with each other to, in a broad sense, enhance their capabilities. The end goal is to enlarge the design space of SC converters, as shown in Fig. 1.5, or even enable entirely new types of SC converters that can handle large VCR ranges. This, in turn, does not only improve the performance for existing use cases, but also enables integrated SC converters for new applications.

1.3   Outline

This book is organized as follows. Chapter 2 gives an introduction to the fundamental principles behind monolithic SC converters. The basic output impedance model of this type of converter is explained, together with other loss factors that are important in the monolithic context. With this model, their optimization is done at a high level of abstraction and the high- and low power-density regimes are discussed in more detail. Moreover, several common topology families are compared with one another under different assumptions regarding the devices. Based on this theoretical background, two figures of merits are introduced, while a third is suggested for a PDN-related context. The theoretical basis is expanded with the presentation of a voltage-domain based analysis method in Chap. 3. Using this analysis a fundamental law of conventional SC DC–DC converters is proven, which is then in turn used to demonstrate that multiphase converters have a fundamental edge over their two-phase cousins in the low power-density regime.

A first advanced multiphasing technique is introduced in Chap. 4 that primarily focuses on low power-densities. Here, the efficiency boundaries are pushed beyond what was previously deemed possible in the monolithic context, by continuously recycling parasitic charge between converter cores. A new fundamental efficiency limit is derived by including transistor leakage losses in an updated model. As part of this technique, multiple DC voltages are generated, which are further explored in Chap. 5 as a means to power internal circuitry within a main SC converter.

Chapter 6 shifts the attention to the limited capacitance density on-chip. Two additional advanced multiphasing techniques are proposed that spread charge transfers between flying capacitors out over time, leading to lower charge-sharing losses or, alternatively, higher effective flying capacitance. The techniques are shown to be compatible with multiple topologies, though especially with large-VCR Dickson converters.

A fundamentally new type of SC converter with large voltage swing capacitors is introduced in Chap. 7. Enabled by advanced multiphasing, this topology breaks the strong connection between VCR, output power, and efficiency that limits conventional SC converters, and is found to instead behave like a gyrator. As a result, the topology is the first purely capacitive DC–DC converter topology that can efficiently and continuously scale its VCR.

Finally, the general conclusions of this book, together with the major contributions are discussed in Chap. 8, in addition to promising future research directions.

Chapter 2
Fully Integrated Switched-Capacitor Fundamentals

As discussed in Chap. 1, there is a clear drive towards monolithic power conversion, and SC converters in particular are an excellent candidate for full integration. Before proceeding to more advanced topics, this chapter will go over the fundamental principles behind SC DC–DC converters. The first part of this chapter deals with the working principle of a switched-capacitor converter. Next, an overview is given of the SC output impedance model and the different loss elements in the monolithic context. Using this model, the optimization of fully integrated switched-capacitors is discussed for different power-density regimes, and device assumptions. Finally, a number of figures of merit are introduced that will be used in the later chapters of this work.

2.1 Working Principle

2.1.1 Charge-Sharing Losses

Consider the circuit set-up in Fig. 2.1. Here, there is a capacitor with capacitance C, that has a certain starting voltage, V_C. When the ideal switch closes at $t = 0$, the capacitor will connect to a terminal with a DC voltage of $V_C + \Delta V$, and will be charged to this voltage. At $t > 0$, the energy in this system has changed as follows:

$$\Delta E = C \frac{(V_C + \Delta V)^2 - V_C^2}{2}. \tag{2.1}$$

Taken into account the energy taken from the voltage supply,

$$E_{supply} = C \Delta V (V_C + \Delta V), \tag{2.2}$$

© Springer Nature Switzerland AG 2020
N. Butzen, M. Steyaert, *Advanced Multiphasing Switched-Capacitor DC-DC Converters*, https://doi.org/10.1007/978-3-030-38735-8_2

Fig. 2.1 A capacitor with
initial voltage, V_C, being
charged to a voltage
$V_C + \Delta V$ at time $t = 0$

Fig. 2.2 A two-phase 2:1 SC
converter topology

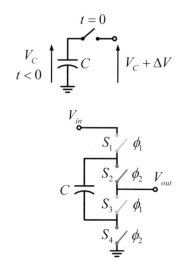

the lost energy of this operation, E_{cs}, can then be calculated:

$$E_{cs} = C \frac{\Delta V^2}{2}. \qquad (2.3)$$

This loss is typically referred to as *charge-sharing loss* and is inherent to charging
and discharging capacitors using voltage sources or other capacitors. Charge-sharing
plays a central role in the design of SC converters. Because the amount of energy
lost depends on the voltage difference between the start- and end voltage squared,
the general goal of SC converters is to minimize this voltage difference as much as
possible.

2.1.2 Topology

SC converters convert one voltage to another by constantly changing the connections
among their capacitors, and between their capacitors and the external nodes
(typically V_{in}, V_{out}, and V_{ss}) over time. Each distinct set of connections that occurs
simultaneously is a *phase*, and a set of phases in a specific order is referred to as a
topology. A topology is thus, in practice, realized by activating and deactivating a
set of switches between the outer terminals and the capacitors at the proper time.

Figure 2.2 portrays an example SC converter topology. This particular topology
has two distinct phases, ϕ_1 and ϕ_2. In the first phase, switches S_1 and S_3 are
conducting, while in the second phase, S_2 and S_4 connect the capacitor's top- and
bottom-plate (TP and BP) to V_{out} and the ground, respectively. To make it easier
to understand the working principles of a topology, it is not unusual to look at the

topology's phase diagram, also shown in Fig. 2.3. In this representation, the switches are omitted, and instead the configuration of the capacitors is shown in each phase.

Because of charge-sharing losses, it is presumed that within a SC topology, the capacitors ideally have a constant voltage across their terminals when the converter is unloaded. This way, the converter will only have charge-sharing losses if there is a load at the output. However, this condition also limits the topology's VCR to a small range near a fixed rational value, the topology's ideal VCR (iVCR). For most topologies, the iVCR can easily be deduced from a phase diagram. As an example, from Fig. 2.3 it can be appreciated that the capacitor's voltage will be set to $V_{in} - V_{out}$ in phase one, and V_{out} in phase two. If the capacitor's voltage does not change over time, then it follows that the iVCR must be 2:1.

The number of capacitors and phases in a topology determines the iVCRs that can be achieved. For two-phase converters, an iVCR= P/Q, where P and Q are strictly positive integers that are coprime, can be realized if and only if

$$P, Q \leq F_{k+2}, \tag{2.4}$$

with F_i the ith Fibonacci number, and k the number of capacitors [MM95]. For p-phase SC converters (2.4) can be further generalized to

$$P, Q \leq F_{k+p}^{(p)}, \tag{2.5}$$

where $F_i^{(p)}$ is the ith Fibonacci number of pth order. The largest realizable iVCRs for a selection of k and p are listed in Table 2.1. As can be seen, by increasing the number of phases, the maximum iVCRs tend towards powers of two.

Fig. 2.3 Phase diagram of a two-phase 2:1 SC converter topology

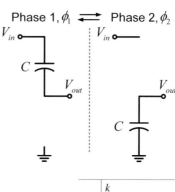

Table 2.1 Largest achievable iVCR with p phases and k capacitors

	k			
p	1	2	3	4
2	2	3	5	8
3	2	4	7	13
4	2	4	8	15
∞	2	4	8	16

Given a desired iVCR, there are multiple topologies that can be used. Each topology has different characteristics that can make it more or less attractive for a specific situation. In addition, some families of topologies have been identified that are constructed similarly. Notable examples include the Series-Parallel- [BS11a], the Dickson [Dic76], the Fibonacci [UIOH91], the Ladder- and the successive approximation register (SAR) converter [BWG$^+$13].

2.2   Modeling

2.2.1   Output Impedance

Switched-capacitor converters can be modeled using a combination of an ideal transformer with a finite output impedance [MM95], as portrayed in Fig. 2.4. This output impedance, R_{out}, has a behavior that depends on the converter's switch frequency, f_{sw}. From Fig. 2.5 it can be appreciated that this behavior has two distinct asymptotic limits.

Fig. 2.4 Switched-capacitor output resistance model

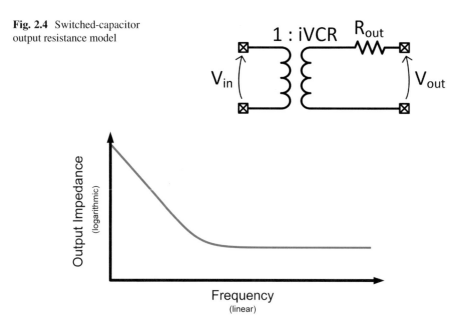

Fig. 2.5 Output impedance of a SC converter

Slow-Switching Limit

In the slow-switching limit (SSL), the converter's capacitors are assumed to charge and discharge fully [See09]. This implies that the time constants of the capacitor (dis)charges are insignificantly relative to the phase duration.

To determine the converter's output impedance in SSL, R_{SSL}, one must relate the charge-sharing losses of each capacitor, C_i, in each phase to the total amount of charge transferred to the output over a full converter period, q_{out}. This is achieved using charge multiplier elements, $a_{c,ij} = q_{c,ij}/q_{out}$, where $q_{c,ij}$ is the charge transferred by capacitor i in phase j [See09]. Using (2.3), the total SSL charge sharing power loss, P_{SSL} can then be calculated:

$$P_{SSL} = \frac{f_{sw}q_{out}^2}{2} \sum_{caps,i} \sum_{phases,j} \frac{a_{c,ij}^2}{C_i}. \tag{2.6}$$

Given that $P_{SSL} = R_{SSL}I_{out}^2$, and that the output current, $I_{out} = f_{sw}q_{out}$, the SSL impedance is obtained [See09]

$$R_{SSL} = \sum_{caps,i} \sum_{phases,j} \frac{a_{c,ij}^2}{2f_{sw}C_i}. \tag{2.7}$$

Note that the charge multiplier elements of a single capacitor should add up to zero if the converter is in steady-state. Because of this, (2.7) can be simplified for two-phase converters:

$$R_{SSL} = \sum_{caps,i} \frac{a_{c,i}^2}{f_{sw}C_i}, \tag{2.8}$$

where $a_{c,i} = q_{c,i}/q_{out}$ and $q_{c,i}$ is the charge transferred by capacitor i in either phase.

As can be appreciated from (2.7) and (2.8), the SSL behavior is determined largely by the charge multiplier elements, which in turn depend on the topology. These elements can be calculated using either Kirchhoff's current law (KCL) on each phase, or just by visual inspection using a topology's phase diagram. Moreover, these elements will also determine the optimal (relative) sizing of the capacitors. To emphasize this fact, it is possible to separate the topological influence on the SSL from all other factors [See09, BS11b]. For designs using a single, voltage-independent capacitor technology, one can consider a total flying capacitance, C_{tot}, being divided into smaller capacitances using relative capacitor sizings, $C_{\%,i}$. From this, a topological parameter, K_C, can be defined as follows:

$$K_C = \sum_{caps,i} \sum_{phases,j} \frac{a_{c,ij}^2}{2C_{\%,i}}, \tag{2.9}$$

which in turn allows one to rewrite (2.7):

$$R_{SSL} = \frac{K_C}{f_{sw}C_{tot}}. \tag{2.10}$$

Fast-Switching Limit

When the charge transfers within the SC converter are severely limited by the switches' resistance, such that the currents going through the capacitors and transistors are approximately constant, the SC converter is said to be in the fast-switching limit (FSL). In this regime, the switches will be the determining factor of the FSL output impedance, R_{FSL}. This impedance can be similarly calculated by considering the loss contribution, $P_{loss,ij}$, of every transistor i in every phase j:

$$P_{loss,ij} = \frac{D_j I_{R,ij}^2}{G_i}, \tag{2.11}$$

with D_j the duty cycle of phase j, G_i the conductance of switch i, and $I_{R,ij}$ the current that flows through switch i in phase j. Now, using the switch charge multiplier elements, $a_{R,ij} = q_{R,ij}/q_{out}$, where $q_{R,ij}$ is the charge that is transferred through switch i during phase j, and considering that $f_{sw}q_{R,ij} = D_j I_{R,ij}$, the general formula for the FSL impedance can be deduced [See09]:

$$R_{FSL} = \sum_{switches,i} \sum_{phases,j} \frac{a_{R,ij}^2}{D_j G_i}. \tag{2.12}$$

For two-phase converters with equal phase duration, (2.12) can be further simplified:

$$R_{FSL} = 2 \sum_{switches,i} \frac{a_{R,i}^2}{G_i}, \tag{2.13}$$

where $a_{R,i}$ is the charge multiplier element of switch i in the conducting phase.

As with the SSL regime, the topological influence can be extracted from the FSL impedance. Assuming a single, voltage-independent switch technology is used, K_S can be defined as

$$K_S = \sum_{switches,i} \sum_{phases,j} \frac{a_{R,ij}^2}{D_j G_{\%,i}}, \tag{2.14}$$

with $G_{\%,i} = G_i/G_{tot}$ and $G_{tot} = \sum_{switches,i} G_i$, the total switch conductance. This in turn simplifies (2.12) to

$$R_{FSL} = \frac{K_S}{G_{tot}}. \tag{2.15}$$

Practical Output Impedance

Both the SSL- and the FSL-impedance only describe the converter's behavior accurately in extreme cases. In a well-optimized design, though, the SC converter's charge transfers will not complete fully, but will still be exponential in nature. For some topologies, like the 2:1 converter, it is possible to come up with an analytical closed-form solution to describe this in-between behavior [PSS16]. In all other cases, it can be approximated using a combination of the asymptotic limits. The most common approach is to take a Euclidean norm of the SSL- and FSL-impedance:

$$R_{out} \approx \sqrt{R_{FSL}^2 + R_{SSL}^2}. \tag{2.16}$$

It is also possible to use a general p-norm where the p factor is carefully selected [Mak12]. While both of these approaches are reasonably accurate, their use does significantly complicate the optimization and, by extension, the analysis of SC converters. Therefore, this chapter will instead use a simple summation, as was suggested by [LSA11]:

$$R_{out} \approx R_{FSL} + R_{SSL}. \tag{2.17}$$

2.2.2   Extrinsic Losses

Next to the losses attributed to the finite output impedance of the converter, there are also a number of losses that are present in any practical SC converter implementation. These losses are considered to be external to the output impedance model discussed in the previous section, and are thus referred to as *extrinsic*.

Parasitic Coupling

In the output impedance model, there is no inherent downside to switching a capacitor between different voltage domains. A practical capacitor, though, will have some parasitic coupling on both terminals that will be charged or discharged in such cases. In the monolithic environment, which is inherently planar, the parasitic capacitor is usually dominated by the coupling of the capacitor terminals to the substrate. Often, this parasitic capacitor is referred to as the *bottom-plate* (BP) capacitor, because historically the bottom terminal was the one closest to the substrate, and thus had the highest parasitic coupling. That being said, BP can also refer to the terminal with the lowest voltage, relative to the *top-plate* (TP).

The losses related to the parasitic coupling, P_{par}, are also charge-sharing losses. As such, according to (2.3), it is the voltage swing across each parasitic capacitor, $C_{par,i}$, at the beginning of each phase j that plays a significant role:

$$P_{par} = \sum_{caps,i} \sum_{phases,j} \frac{f_{sw} C_{par,i} \Delta V_{ij}^2}{2}. \tag{2.18}$$

Because these voltage swings are, of course, determined by the topology, it is desirable to separate their impact from the other design parameters. When using a single-capacitor implementation, the ratio between a capacitor's parasitic coupling and the flying capacitance itself, α_{par}, is approximately fixed. In fact, only at very small capacitor sizes, the parasitic fringe-capacitance has a notable influence, thereby causing α_{par} to increase. Using this approximation, the parasitic coupling topology factor, K_{par}, can be defined:

$$K_{par} = \sum_{caps,i} \sum_{phases,j} \frac{C_{\%,i} \Delta v_{ij}^2}{2}, \tag{2.19}$$

where $\Delta v_{ij} = \Delta V_{ij}/V_{out}$ is the normalized parasitic voltage swing. In turn, (2.18) can be reformulated to

$$P_{par} = \alpha_{par} f_{sw} C_{tot} V_{out}^2 K_{par}. \tag{2.20}$$

Transistor Driving Losses

In addition to the capacitor's parasitic coupling losses, there is also an energy cost to activating any practical implementation of switches. When using transistors in a CMOS technology, this energy cost is mainly related to charging the transistor's gate capacitance. While there are also losses related to the drain-source capacitance, these are usually either neglected or included to the gate capacitance [LSA11,

SS15a]. Other related dynamic losses (e.g. gate driver) can be included in similar fashion.

Because the gate capacitor is non-linear, the related losses, P_{trans}, can be expressed in terms of the gate charge, or even in terms of a general activation energy, $E_{on,i}$:

$$P_{trans} = f_{sw} \sum_{switches,i} b_i E_{on,i}, \qquad (2.21)$$

with b_i the number of activations (and deactivations) per cycle for switch i.

Both the conductance and the total gate charge scale linearly with the transistor width. As such, assuming only a single type of voltage-independent transistor is used, it is possible to relate (2.21) to the switch conduction:

$$P_{trans} = E_{on} R_{on} f_{sw} G_{tot} \sum_{switches,i} b_i G_{\%,i}, \qquad (2.22)$$

where $E_{on} R_{on}$ is the transistor technology's activation energy per unit conduction. Because the number of activations per switch depends on the topology, it is once again possible to define a topological factor, K_{trans}:

$$K_{trans} = \sum_{switches,i} b_i G_{\%,i}, \qquad (2.23)$$

which in turn simplifies (2.22):

$$P_{trans} = E_{on} R_{on} f_{sw} G_{tot} K_{trans}. \qquad (2.24)$$

Note that for a two-phase converter, or broadly speaking for topologies where every switch activates and deactivates exactly once, K_{trans} is one.

2.3 Optimization

In the previous two sections, a total of four loss mechanisms of fully integrated switched-capacitor converters were identified. Moreover, in each of these loss contributions, the topological influence was separated from the essential design-, and technological parameters.

Generally speaking, the goal of optimization is to maximize the performance of the converter given a certain set of specifications. What metric, or set of metrics, determine the performance, depends heavily on the application for which the converter will be used. For example, for DC/DC step-up converters used in thermal energy scavenging, the lowest input voltage for which the converter still starts-up by itself is critical, and should be minimized [RC10]. That being said, there are still some metrics which are important for nearly all applications: Cost and energy loss. With this in mind, this section will remain agnostic to the application, and focus on optimizing the converter for efficiency versus output power and area. In contrast with previous work, this optimization will first take place on the converter level, followed by the topology level.

2.3.1 Converter Optimization

Combining all losses together, and using (2.17), the total converter losses, P_{loss} are defined as:

$$P_{loss} = (R_{SSL} + R_{FSL})I_{out}^2 + P_{par} + P_{trans}. \tag{2.25}$$

In an additional step, it is useful to normalize these losses by dividing them with the converter output power, P_{out}:

$$P_N = \frac{P_{loss}}{P_{out}}. \tag{2.26}$$

The resulting normalized losses, P_N, can also be directly tied to the converter efficiency, η.

$$\eta = \frac{1}{1 + P_N}. \tag{2.27}$$

From (2.10), (2.15), (2.20), (2.24), and (2.26), it is clear that there are a total of 5 design variables in the expression of the normalized losses: C_{tot}, G_{tot}, f_{sw}, I_{out}, and V_{out}. In this optimization, it is presumed that the output voltage is set by the specifications. Furthermore, because the area of the converter is usually dominated by that of the capacitors, the impact of the switch area on the power-density, $P_D = P_{out}/A_{conv}$, is neglected. Under these assumptions, the normalized losses can be rewritten:

$$P_N = \frac{I_{out}^2}{P_{out}} \frac{K_C}{f_{sw} C_{tot}} + \frac{I_{out}^2}{P_{out}} \frac{K_S}{G_{tot}} + \frac{V_{out}^2}{P_{out}} \alpha_{par} C_{tot} f_{sw} K_{par}$$

$$+ \frac{E_{on} R_{on}}{P_{out}} f_{sw} G_{tot} K_{trans},$$

$$P_N = \frac{P_D}{V_{out}^2} \frac{K_C}{f_{sw} \alpha_C} + \frac{P_D}{V_{out}^2} \frac{K_S}{G_{tot}'} + \frac{V_{out}^2}{P_D} \alpha_{par} \alpha_C f_{sw} K_{par}$$

$$+ \frac{E_{on} R_{on}}{P_D} f_{sw} G_{tot}' K_{trans}, \tag{2.28}$$

where α_C is the capacitor technology's capacitance density, and G'_{tot} is the total switch conductance, both per unit area. Now the converter losses can be optimized for the remaining three parameters, G'_{tot}, f_{sw}, and P_D:

$$\frac{\partial P_N}{\partial f_{sw}} = 0 \Leftrightarrow P_{SSL} = P_{par} + P_{trans}, \tag{2.29}$$

$$\frac{\partial P_N}{\partial G'_{tot}} = 0 \Leftrightarrow P_{FSL} = P_{trans}, \tag{2.30}$$

$$\frac{\partial P_N}{\partial P_D} = 0 \Leftrightarrow P_{FSL} + P_{SSL} = P_{par} + P_{trans}. \tag{2.31}$$

Each of the equations above gives a straight-forward condition for the optimal frequency, switch conductance, and power-density respectively. From (2.29), it is clear that the optimal frequency is such that the SSL losses match the parasitic coupling- and transistor driving losses, while (2.29) dictates that when the FSL losses are equal to the transistor driving losses, the switch conductance is optimal. Regardless of the desired power-density, (2.29) and (2.30) need to be met. Combining both together, it can be deduced that, in an optimal design, the total conduction losses, $P_{SSL} + P_{FSL}$, must be larger than the total extrinsic losses, $P_{par} + P_{trans}$. Furthermore, if power-density can be chosen freely, the highest efficiency is obtained when P_{trans} tends to zero, but this also means that in turn P_{FSL} and the power-density itself tend to zero. At this point the optimal frequency leads to P_{SSL} being equal to P_{par}, and the corresponding normalized losses, $P_{N,min}$, are

$$P_{N,min} = \frac{2\sqrt{P_{SSL} P_{par}}}{P_{out}}, \tag{2.32}$$

$$P_{N,min} = 2\sqrt{\alpha_{par} K_{par} K_c}. \tag{2.33}$$

Thus, as discussed in Chap. 1, at low power-densities an effective efficiency ceiling is formed, which depends on the technology specific parasitic coupling parameter, α_{par}, and the chosen topology.

To figure out what happens at larger power-densities, one has to look at how the optimal frequency, $f_{sw,opt}$, and switch conductance, $G'_{tot,opt}$, scale with power-density. Rewriting (2.29) and (2.30) yields

$$f_{sw,opt} = P_D \sqrt{\frac{K_C}{\alpha_C V_{out}^2} \frac{1}{\alpha_{par} \alpha_C V_{out}^2 K_{par} + E_{on} R_{on} K_{trans} G'_{tot,opt}}}, \tag{2.34}$$

$$G'_{tot,opt} = P_D \sqrt{\frac{K_S}{V_{out}^2} \frac{1}{E_{on} R_{on} K_{trans} f_{sw,opt}}}. \tag{2.35}$$

Formula (2.34) reveals that for sufficiently small switch conductances, the optimal frequency scales approximately linearly with power-density. The optimal conductance, in contrast, scales linearly with power-density and is inversely proportionate to the square root of the frequency. The resulting net scaling is thus proportionate to $\sqrt{P_D}$ for low switch conductances. However, with increasing power-densities, the switch conductance will eventually reach a level where it can no longer be neglected in the optimal frequency calculation, and thus the scaling of both parameters is altered as shown in Fig. 2.6a. This critical point occurs when $P_{trans} = P_{par}$. For power-densities much larger than this turning-point, P_{par} can even be neglected. Combining this with (2.29) and (2.30), the high power-density region can be categorized as the region where $P_{SSL} \approx P_{FSL} = P_{trans}$. Here, a clear trade-off between the normalized losses, $P_{N,HD}$, and power-density emerges:

$$P_{N,HD} \approx 3\frac{\sqrt[3]{P_{SSL}P_{FSL}P_{trans}}}{P_{out}}, \tag{2.36}$$

$$P_{N,HD} \approx 3\sqrt[3]{P_D K_C K_S K_{trans}\frac{E_{on}R_{on}}{\alpha_C V_{out}^4}}. \tag{2.37}$$

Figure 2.6 gives a visual summary of the previous discussion. At lower power-densities, the losses are dominated by the capacitor charge-sharing and parasitic coupling losses, and the normalized losses reach a minimum. Because both losses are related to the capacitors, a converter in this regime is limited by the capacitor technology. For high power-densities, in contrast, transistor losses are an additional factor, and both capacitor- and transistor technology become important. From Fig. 2.6b, c it is clear that in the transitional region, neither of the assumptions made earlier holds. While it is possible to approximate the normalized losses in this region, this does not provide any additional insights [SLM16].

2.3.2 Topological Optimization

Given a topology, it is not always possible to minimize each topological factor, K_C, K_S, K_{trans}, and K_{par}, separately. For example, both K_C and K_{par} depend on the relative distribution of the flying capacitors, while K_S and K_{trans} similarly rely on the switch distribution. Despite these co-dependencies, a popular approach is to only optimize the capacitor- and transistor distributions for conduction losses [See09]. In this discussion, the derived equations for the normalized losses will be used to arrive at improved resource distributions.

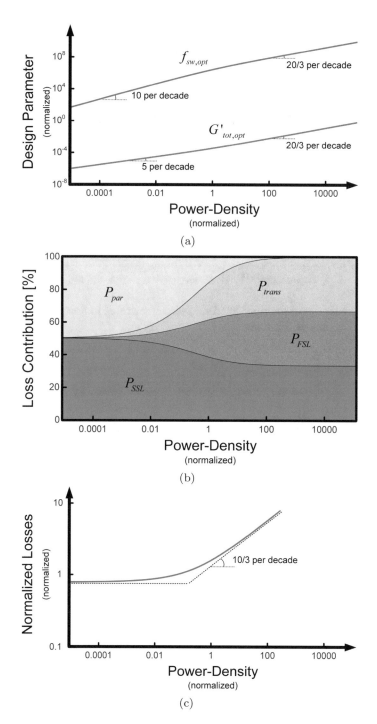

Fig. 2.6 Scaling of (**a**) frequency and switch conductance per unit area, (**b**) relative loss contribution, and (**c**) normalized losses of an optimized fully integrated switched-capacitor converter versus power-density

Low Power-Density

From (2.33), it follows that at low power-densities, the normalized losses are proportionate to $\sqrt{K_C K_{par}}$. As a result, it is desirable to minimize the capacitor distribution for this factor in this regime. For a converter using a single-capacitor technology that is voltage-independent (or for which the voltage rating is larger than the largest required voltage by the topology), the function to optimize is thus

$$K_C K_{par} = \left(\sum_{caps,i} C_{\%,i}^{-1} \sum_{phases,j} \frac{a_{c,ij}^2}{2} \right) \left(\sum_{caps,i} C_{\%,i} \sum_{phases,j} \frac{\Delta v_{ij}^2}{2} \right). \qquad (2.38)$$

Taking the partial derivative for each variable $C_{\%,k}$ yields a set of equations:

$$C_{\%,k} = \sqrt{\frac{\sum_{phases,j} a_{c,kj}^2}{\sum_{phases,j} \Delta v_{kj}^2}} \sqrt{\frac{\sum_{caps,i \neq k} C_{\%,i} \sum_{phases,j} \Delta v_{ij}^2}{\sum_{caps,i \neq k} C_{\%,i}^{-1} \sum_{phases,j} a_{c,ij}^2}}. \qquad (2.39)$$

The second term of the right-hand side can be rewritten to depend on $C_{\%,k}$ itself, resulting in an alternative form

$$C_{\%,k} = \sqrt{\frac{\sum_{phases,j} a_{c,kj}^2}{\sum_{phases,j} \Delta v_{kj}^2}} \sqrt{\frac{2K_{par,opt} - C_{\%,k} \sum_{phases,j} \Delta v_{kj}^2}{2K_{C,opt} - C_{\%,k}^{-1} \sum_{phases,j} a_{c,kj}^2}}, \qquad (2.40)$$

$$\frac{2K_{C,opt} C_{\%,k}^2}{\sum_{phases,j} a_{c,kj}^2} - C_{\%,k} = \frac{2K_{par,opt}}{\sum_{phases,j} \Delta v_{kj}^2} - C_{\%,k}, \qquad (2.41)$$

$$C_{\%,k} = \sqrt{\frac{\sum_{phases,j} a_{c,kj}^2}{\sum_{phases,j} \Delta v_{kj}^2}} \sqrt{\frac{K_{par,opt}}{K_{C,opt}}}, \qquad (2.42)$$

where $K_{C,opt}$ and $K_{S,opt}$ are the optimal value of K_C and K_S, respectively. Finally, the relative size of $K_{C,opt}$ and $K_{S,opt}$ is chosen such that $\sum_{caps,i} C_{\%,i} = 1$, and the optimal capacitor size is subsequently obtained:

$$C_{\%,k} = \frac{\sqrt{\dfrac{\sum_{phases,j} a_{c,kj}^2}{\sum_{phases,j} \Delta v_{kj}^2}}}{\sum_{caps,i} \sqrt{\dfrac{\sum_{phases,j} a_{c,ij}^2}{\sum_{phases,j} \Delta v_{ij}^2}}}. \qquad (2.43)$$

In other words, each capacitor should be sized proportionate to the Euclidean norm of its charge multiplier vector, and inversely proportionate to same norm of its BP voltage swings. For two-phase converters, this is reduced to $C_{\%,k} \propto |a_{c,k}/\Delta v_k|$, with Δv_k the normalized BP voltage swing. Using optimal sizing, the minimum

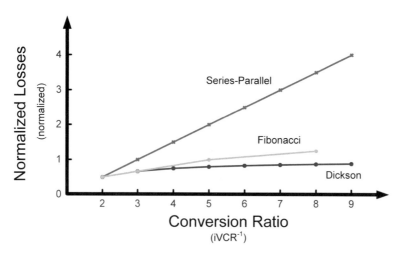

Fig. 2.7 Scaling of normalized losses versus conversion ratio for Series-Parallel, Fibonacci and Dickson converter topology families in the low power-density regime, assuming voltage-independent capacitor implementation

topological value is

$$(K_C K_{par})_{opt} = \frac{1}{4} \left(\sum_{caps,i} \sqrt{\sum_{phases,j} a_{c,ij}^2 \sum_{phases,j} \Delta v_{ij}^2} \right)^2. \qquad (2.44)$$

Note that this result relies on the assumption that each capacitor has a non-zero BP swing. This means that for topologies that require non-flying buffer capacitors, such as the Ladder converter, a different sizing distribution needs to be used.

The corresponding optimal low power-density normalized losses are shown in Fig. 2.7 for three different topology families. It can be appreciated that the Dickson converter has the lowest losses of all compared topologies in this regime, which explains its popularity in the literature [MSBS14, SPSS15]. The Fibonacci converter has larger losses, and, unlike when using off-chip components, its reduced component count relative to the Dickson converter is mostly irrelevant in the monolithic context. The series-parallel converter, finally, has by far the largest losses of the compared topologies and is thus best avoided for low power-densities.

High Power-Density

In the high power-density regime, the focus shifts towards the switches. According to (2.37), the factor to optimize is $K_C K_S K_{trans}$. Because the parasitic coupling losses of the capacitors are neglected, it is possible to optimize K_C separately [See09]:

$$C_{\%,k} = \frac{\sqrt{\sum_{phases,j} a_{c,kj}^2}}{\sum_{caps,i} \sqrt{\sum_{phases,j} a_{c,ij}^2}}, \tag{2.45}$$

$$(K_C)_{opt} = \frac{1}{2} \left(\sum_{caps,i} \sqrt{\sum_{phases,j} a_{c,ij}^2} \right)^2. \tag{2.46}$$

The other parameters, K_S and K_{trans}, are correlated and need to be optimized together. Using a similar strategy to the capacitor optimization in the low power-density regime, the optimal switch sizes are deduced:

$$G_{\%,k} = \frac{\sqrt{b_k^{-1} \sum_{phases,j} D_j^{-1} a_{R,kj}^2}}{\sum_{switches,i} \sqrt{b_i^{-1} \sum_{phases,j} D_j^{-1} a_{R,ij}^2}}. \tag{2.47}$$

Combining (2.46) and (2.47), in turn yields the minimum topological factor

$(K_C K_S K_{trans})_{opt}$

$$= \frac{1}{2} \left(\sum_{caps,i} \sqrt{\sum_{phases,j} a_{c,ij}^2} \sum_{switches,i} \sqrt{b_i \sum_{phases,j} D_j^{-1} a_{R,ij}^2} \right)^2. \tag{2.48}$$

2.3.3   Voltage-Dependent Devices

When converting large voltages relative to the integrated technology's supply voltage, it is usually not possible to only make use of a single-capacitor implementation. In this situation, multiple capacitor types are combined in one converter, or capacitors need to be put in series to arrive at a high-voltage device. Alternatively, some capacitor types, like the Metal-Oxide-Metal (MOM) capacitor, can simply be adjusted to handle larger voltages. Interestingly enough, if this is done without changing the height of the dielectric volume and under constant electric field, then the energy density stays constant, as demonstrated in Fig. 2.8. The same is of course true when putting multiple capacitors in series.

Similarly, when a switch needs to be able to block a voltage larger than the transistors' maximum voltage rating, it is typically implemented using multiple stacked transistors [SPSD05]. Figure 2.9 illustrates how the GV^2 product per unit die area of the switch, α_S, remains constant.

Both capacitors and switches thus have a cost associated with their voltage squared. To take this into account, the earlier optimization can be adjusted to optimize the relative flying capacitor- and switch areas, $A_{C\%i}$ and $A_{R\%i}$, respectively, rather than the relative capacitances and conductances. A flying capacitor's

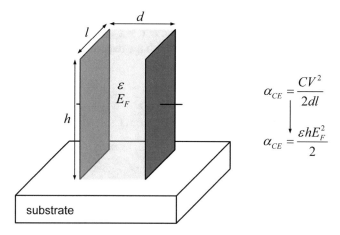

Fig. 2.8 Energy density per unit die area, α_{CE}, of a parallel-plate capacitor, with d the distance between the parallel plates, l and h, respectively, the length and height of the capacitor, ϵ the dielectric's permittivity, and E_F the electric field

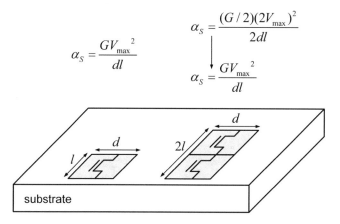

Fig. 2.9 GV^2-product per unit die area, α_S, of a switch consisting of stacked transistors. d and l are, respectively, the width and length of the switch die area, V_{max} is the transistor's maximum voltage, and G the transistor conductance

capacitance and a switch's conductance can be related to their respective parameter as follows:

$$C_i = \frac{2\alpha_{CE}A_{C\%i}}{V_{Ci}^2}A_{C,tot}, \tag{2.49}$$

$$G_i = \frac{\alpha_S A_{R\%i}}{V_{Ri}^2}A_{R,tot}, \tag{2.50}$$

with V_{Ci} the capacitor's bias voltage, V_{Ri} the switch's block voltage when not conducting, $A_{C,tot}$ the total capacitor area, and $A_{R,tot}$ the total switch area. Given the above relationships, the corresponding voltage dependent topological factors are identified:

$$K'_C = \sum_{caps,i} \frac{v_{Ci}^2}{A_{C\%i}} \sum_{phases,j} \frac{a_{c,ij}^2}{2}, \tag{2.51}$$

$$K'_{par} = \sum_{caps,i} A_{C\%i} \sum_{phases,j} \frac{\Delta v_{ij}^2}{2}, \tag{2.52}$$

$$K'_S = \sum_{switches,i} \frac{v_{Ri}^2}{A_{R\%i}} \sum_{phases,j} \frac{a_{R,ij}^2}{D_j}, \tag{2.53}$$

$$K'_{trans} = \sum_{switches,i} b_i A_{R\%i}, \tag{2.54}$$

where $v_{Ci} = V_{Ci}/V_{out}$ and $v_{Ri} = V_{Ri}/V_{out}$ are the normalized capacitor bias- and transistor block voltage, respectively.

The change of topological factors impacts the equations for the normalized losses in the low- and high power-density regime, (2.33) and (2.37). However, using (2.32) and (2.36), the altered solutions can be deduced:

$$P'_{N,min} = V_{out}\sqrt{2K'_C K'_{par} \frac{\alpha_{Cpar}}{\alpha_{CE}}}, \tag{2.55}$$

$$P'_{N,HD} \approx 3\sqrt[3]{P_D K'_C K'_S K'_{trans} \frac{E_{on}}{A_{on}} \frac{1}{\alpha_S \alpha_{CE}}}. \tag{2.56}$$

Here, α_{Cpar} is the capacitor's parasitic coupling density, and E_{on}/A_{on} is the switch activation energy, both per unit area. Comparing (2.55) and (2.56) to (2.33) and (2.37), respectively, it becomes evident that in both regimes, the losses scale better towards lower output voltages: (2.55) is directly proportionate to V_{out}, while (2.56) is independent of V_{out}, rather than proportionate to $V_{out}^{-4/3}$. Naturally, this is only valid until the output voltage drops below the smallest device voltage rating (e.g. the technology's supply voltage), but it does reveal that at some fundamental level, switched-capacitor converters would benefit significantly from being able to finely tune the capacitance- and conduction density according to the required device voltage rating, even well below the technology voltage.

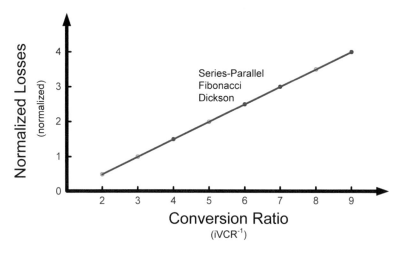

Fig. 2.10 Scaling of normalized losses versus conversion ratio for Series-Parallel, Fibonacci and Dickson converter topology families in the low power-density regime, assuming constant energy-density capacitor implementation

Low Power-Density

For low power-densities, (2.55) reveals that $K'_C K'_{par}$ must be minimized. The result of this optimization is (Fig. 2.10)

$$A_{C\%k} = \frac{\sqrt{\dfrac{v_{Ci}^2 \sum_{phases,j} a_{c,kj}^2}{\sum_{phases,j} \Delta v_{kj}^2}}}{\sum_{caps,i} \sqrt{\dfrac{v_{Ci}^2 \sum_{phases,j} a_{c,ij}^2}{\sum_{phases,j} \Delta v_{ij}^2}}}, \tag{2.57}$$

$$(K'_C K'_{par})_{opt} = \frac{1}{4} \left(\sum_{caps,i} \sqrt{v_{Ci}^2 \sum_{phases,j} a_{c,ij}^2 \sum_{phases,j} \Delta v_{ij}^2} \right)^2. \tag{2.58}$$

The capacitance area should thus be proportional to the capacitor bias voltage, which means that the capacitance itself is inversely proportionate to the voltage. Equation (2.58) can be further simplified for two-phase converters to

$$(K'_C K'_{par})_{opt} = \left(\sum_{caps,i} |v_{Ci} a_{c,i} \Delta v_i| \right)^2. \tag{2.59}$$

Equation (2.59) shows the scaling of these normalized losses versus conversion ratio for a selection of topology families. Surprisingly, the Series-Parallel, the

Fibonacci and the Dickson converter all have the exact same normalized losses for every conversion ratio:

$$(P'_{N,min})_{opt} \propto \frac{1 - iVCR}{2iVCR}. \tag{2.60}$$

It turns out that this remarkable fact is due to an underlying fundamental relationship that all switched-capacitors converter topologies share, which will be proven and discussed in more detail in Chap. 3.

High Power-Density

As was the case with voltage-independent devices, the product to optimize in the high power-density regime is a combination of the SSL, FSL, and transistor-driving topological factors: $K'_C K'_S K'_{par}$. This optimization yields the following result:

$$A_{C\%k} = \frac{\sqrt{v_{Ck}^2 \sum_{phases,j} a_{c,kj}^2}}{\sum_{caps,i} \sqrt{v_{Ci}^2 \sum_{phases,j} a_{c,ij}^2}}, \tag{2.61}$$

$$A_{R\%k} = \frac{\sqrt{b_k^{-1} v_{Rk}^2 \sum_{phases,j} D_j^{-1} a_{R,kj}^2}}{\sum_{switches,i} \sqrt{b_i^{-1} v_{Ri}^2 \sum_{phases,j} D_j^{-1} a_{R,ij}^2}}, \tag{2.62}$$

$$(K_C K_S K_{trans})_{opt}$$

$$= \frac{1}{2} \left(\sum_{caps,i} \sqrt{v_{Ci}^2 \sum_{phases,j} a_{c,ij}^2} \sum_{switches,i} \sqrt{b_i v_{Ri}^2 \sum_{phases,j} D_j^{-1} a_{R,ij}^2} \right)^2. \tag{2.63}$$

The latter can be reduced for two-phase converters to

$$(K_C K_S K_{trans})_{opt} = 2 \left(\sum_{caps,i} |v_{Ci} a_{c,i}| \sum_{switches,i} |v_{Ri} a_{R,i}| \right)^2. \tag{2.64}$$

Several topology families are compared with each other at high power-density in Fig. 2.11. Overall, the Ladder converter shows the worst performance out of the four, having approximately 74% higher losses compared with the Series-Parallel topology for a 5:1 iVCR. The Dickson- and Fibonacci converter both have lower normalized losses than the Ladder topology, and are very close to each other. The

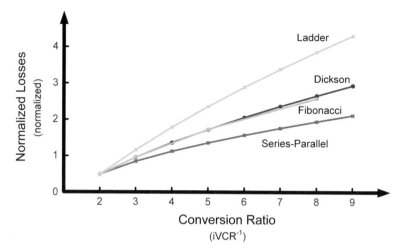

Fig. 2.11 Scaling of normalized losses versus conversion ratio for Series-Parallel, Fibonacci, Dickson and Ladder converter topology families in the high power-density regime, assuming constant energy-density capacitor-, and GV^2-product switch implementation

best performing converter over the entire iVCR range, though, is the Series-Parallel converter.

2.4 Figures of Merit

Figures of merit (FoMs) reduce a complex system to a single figure. A lot of information consequently gets lost, some of which might even be very important for a certain application. That being said, there are good reasons for their use. Ideally, figures of merit are not just a random selection of performance metrics, but are instead based on theory. This does not only enable fair comparison between different designs, but it also learns us something about the fundamental trade-offs during the design itself, or can tell us how useful a design might be in a specific context.

In this section, three figure of merits will be introduced. Some are based on the derivations of Sect. 2.3, others are with a certain context in mind.

2.4.1 Power Delivery Network

As pointed out in Sect. 1.1.2, one of the key performance benefits to integrating a power converter onto the same die as the application is the reduction of losses related to the power delivery network. The PDN FoM derived below attempts to quantify the benefit of integrating a specific design. It is defined as the lowest normalized

PDN losses for which the addition of the fully integrated DC–DC converter still improves the overall system efficiency. In other words, the smaller the FoM, the more suited the converter is for this situation.

Revisiting Fig. 1.2, it is clear that without an integrated converter, there are a certain amount of PDN-induced losses, $P_{PDN}^{(1)}$ which are proportionate to the current drawn by the load, I_{load}, and normalized by the total load power. If an integrated converter is introduced, the PDN losses will be reduced by the converter's current conversion ratio:

$$P_{PDN}^{(2)} = P_{PDN}^{(1)} \frac{I_{in}}{I_{load}}, \tag{2.65}$$

where I_{in} is the input current of the converter. Of course, the converter also adds its own normalized losses, P_N. The break-even point between both determines the PDN FoM:

$$P_{PDN}^{(1)} = P_N + P_{PDN}^{(1)} \frac{I_{in}}{I_{load}} \tag{2.66}$$

$$FoM_{PDN} = \frac{1 - \eta}{\eta - VCR}. \tag{2.67}$$

As a side note, the PDN FoM can also be written using the efficiency enhancement factor (EEF) introduced in [SVBM$^+$11], which compares the input power of a DC–DC converter to that of a linear dropout (LDO) regulator with the same VCR.

$$FoM_{PDN} = \frac{P_N}{EEF}. \tag{2.68}$$

2.4.2 Low Power-Density

In Sect. 2.3.3 it was determined that in the low power-density region, the Series-Parallel-, the Fibonacci- and the Dickson converter all had identical optimized normalized losses when assuming constant energy-density capacitors. In Chap. 3 it will be shown that their performance is, in fact, a fundamental limit for two-phase converters. The proposed low-density FoM simply compares the performance of a converter to this fundamental limit.

Combining (2.55) and (2.60), the fundamental limit is determined to be

$$P_{N,min} = V_{out} \left| \frac{1 - iVCR}{iVCR} \right| \sqrt{\frac{\alpha_{Cpar}}{2\alpha_{CE}}}, \tag{2.69}$$

which in turn leads to the proposed FoM

$$FoM_{LD} = \frac{P_N}{V_{out}} \left| \frac{iVCR}{1 - iVCR} \right|. \tag{2.70}$$

Note that the optimal losses depend on the topology's iVCR, rather than the actual VCR, V_{out}/V_{in}. If there is current drawn from the output of the converter, there will of course be a difference between both. Also, it is important to point out that the lower the FoM, the better the performance of the converter.

2.4.3 High Power-Density

For converters with a high output power-density, (2.56) demonstrates that its normalized losses are proportionate to the topology's topological factors, K'_C, K'_S and K'_{trans}, and the converter's power-density, P_D, as follows [LSA11]:

$$P_N \propto \sqrt[3]{K'_C K'_S K'_{trans} P_D}. \tag{2.71}$$

Even though the Series-Parallel converter was found to have the lowest losses in the comparison of Sect. 2.3.3, there is no guarantee that this is in fact a theoretical optimum. In [See09], however, some limits on the capacitor- and switch- metrics individually were identified, based on the work of [Wol72]:

$$(K'_S)_{opt} \propto (1 - iVCR)^2, \tag{2.72}$$

$$(K'_C)_{opt} \propto (1 - iVCR)^2. \tag{2.73}$$

Considering that for a two-phase converter $K'_{trans} = 1$, combining above equations leads to

$$P_N \propto \sqrt[3]{(1 - iVCR)^4 P_D}. \tag{2.74}$$

The high-density FoM, FoM_{HD}, is then defined as the ratio of the converter's normalized losses to the right-hand side.

$$FoM_{HD} = \frac{P_N}{\sqrt[3]{(1 - iVCR)^4 P_D}}. \tag{2.75}$$

2.5 Conclusion

This chapter dealt with the fundamentals of fully integrated switched-capacitor DC–DC conversion. Charge-sharing losses, a key loss factor associated with charging and discharging capacitors, were shown to play a central role in SC converters. In fact, it was pointed out that, in minimizing these losses, SC topologies are inherently limited by a rational ideal voltage conversion ratio. The basic output impedance model was explained with its two asymptotic behaviors: the slow- and fast switching limit. Transistor driving- and parasitic coupling losses were additionally introduced as important in the monolithic context.

Combining all loss factors together, an optimization was done at the converter level, which revealed the existence of a low- and high power-density regime. In the former, the converter is limited by the capacitor quality, while in the latter, the transistor quality also becomes a contributor. Next, losses were minimized at the topology-level by looking at the ideal relative switch- and capacitor size, both considering voltage-dependent- and independent devices. The Dickson converter was found to be the best candidate out of all compared topology families for low power-density converters with voltage-independent capacitors. The Series-Parallel converter, on the other hand, had the best relative performance at high power-densities using voltage dependent components. Furthermore, using constant energy-density capacitors, the performance of the Series-Parallel-, the Fibonacci- and the Dickson topologies was found to be identical at low power-densities.

In the last part, a total of three figures of merit were introduced. The low- and high-density FoMs compare the performance of a fully integrated SC DC–DC converter to theoretical limits, while the power delivery network FoM evaluates a given converter for their potential to improve the overall system efficiency by reducing PDN-related losses.

Chapter 3
Voltage-Domain Analysis

In Chap. 2, it was pointed out that several topology families share a relationship between their iVCR, and a combination of their capacitor's BP voltage swing, bias voltage and charge multiplier elements. This chapter will introduce a new type of analysis method for power converters, called a voltage-domain analysis (VDA). Using this analysis, it will be shown that a variation on said relationship is in fact shared by all conventional switched-capacitor DC–DC converters, and can thus be considered a general law for this type of converter. This law reveals the underlying trade-offs between different SC topologies, and demonstrated how some converters can be inefficient from a purely topological point of view. In addition, the proposed VDA is also shown to be useful as a visual method for SC topology synthesis.

Parts of this chapter were previously presented at the Nineteenth IEEE Workshop on Control and Modeling for Power Electronics (COMPEL) 2018 in Padua, Italy [BS18a].

3.1 Background

Most fundamental knowledge of switched-capacitor DC–DC converters, or power converters in general, is derived using graph theory. A good example hereof is the pioneering work of Wolaver [Wol69, Wol72], based in turn on the earlier contribution by Moore and Wilson [MW66], that established multiple constraints on a DC–DC converter's passives' and switches' voltages and currents, depending on the voltage conversion ratio. A selection of these constraints, translated for two-phase capacitive step-down converters, are outlined below. While these relationships were critical in determining the performance limits of SC converters, they have seen little other use in the SC literature. Moreover, because they are constraints rather than relations in equality, they do not tell us anything about the trade-offs between different topologies.

© Springer Nature Switzerland AG 2020
N. Butzen, M. Steyaert, *Advanced Multiphasing Switched-Capacitor DC-DC Converters*, https://doi.org/10.1007/978-3-030-38735-8_3

$$\sum_{caps,i} |v_{Ci}| \geq \frac{1 - iVCR}{iVCR} \tag{3.1}$$

$$\sum_{caps,i} |a_{c,i}| \geq 1 - iVCR \tag{3.2}$$

$$\sum_{caps,i} |v_{Ci}a_{c,i}| \geq 1 - iVCR \tag{3.3}$$

Another work, published in 1995, used graph theory to relate the required number of capacitors to achieve a certain VCR to the Fibonacci sequence [MM95], as discussed in Sect. 2.1.2. Said work also determined how many switches it would take to realize this VCR. Since then, however, theoretical work has shifted away from fundamental topology theory towards loss- and equivalent output impedance modeling [PSS16], or switched-capacitor topology synthesis [Kar15, MP17].

This chapter is structured as follows. In Sect. 3.2, the voltage-domain analysis will be introduced, which will in turn be used in Sect. 3.3 to proof a general law for conventional switched-capacitor DC–DC power conversion. Section 3.4 deals with some of the consequences of this new analysis method and law on the design and theory of SC converters. Finally, Sect. 3.5 summarizes the important conclusions of this work.

3.2   A New Analysis Tool

The basic idea of the voltage-domain analysis is to look at the distribution of a device's terminal currents in the voltage domain. Figure 3.1 portrays a general n-terminal device. Its terminals are numbered for ascending terminal voltages, V_1 to V_n, and respective currents, I_1 to I_n. These terminal currents are defined as the time-averaged current and are considered positive by convention when flowing into the device. In addition, it is assumed that all terminals are infinitely decoupled, such that the terminal voltages are perfect DC.

Fig. 3.1 General n-terminal device

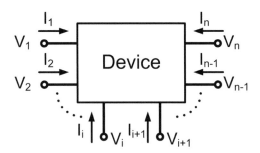

The current distribution of said device in the voltage domain, $I(v)$, can be described as

$$I(v) = \sum_{terminals,i} I_i \delta(v - V_i), \tag{3.4}$$

with $\delta(v)$ the Dirac delta distribution. It is important to point out that the delta distribution is used for its mathematical properties, and not necessarily as a reflection of a particular physical reality. Figure 3.2a shows an example current distribution for an n-terminal device. Note that the y-axis in this plot shows the weights of the $\delta(v)$s. From this current distribution, two more distributions are defined; the cumulative current distribution, $I^*(v)$, and the cumulative power distribution, $P^*(v)$:

$$I^*(v) = \int_{-\infty}^{v} I(\tau)d\tau, \tag{3.5}$$

$$P^*(v) = \int_{-\infty}^{v} I^*(\tau)d\tau. \tag{3.6}$$

Consequently, $I^*(v)$ is a step-function, while $P^*(v)$ is piece-wise linear. Figure 3.2b, c visualize a cumulative current- and power distribution that matches the current distribution of Fig. 3.2a. In these illustrations, $I_i^* = \sum_{j=1}^{i} I_j$, and $P_F^* = P^*(V_n)$.

Naturally, for any device the terminal currents must sum to zero. With this in mind, $I^*(v) = 0$ for any voltage larger than V_n, which implies that $P^*(v)$ is constant for any voltage larger than V_n as well. Furthermore, P_F^*, the value of $P^*(v)$ at the highest terminal voltage, V_n, is equal to the negation of the net power loss of the device:

$$P_F^* = \sum_{i=1}^{n-1} (V_{i+1} - V_i)I_i^*$$

$$= V_n I_{n-1}^* - V_1 I_1^* + \sum_{i=2}^{n-1} V_i(I_{i-1}^* - I_i^*)$$

$$P_F^* = - \sum_{terminals} V_i I_i. \tag{3.7}$$

For a lossless device, P_F^* is consequently zero. As a result, as long as the device in question has finitely many terminals with finite terminal currents, the area of the cumulative power distribution relative to the x-axis is finite as well.

Now, the argument that can be made is that, if a device consists of several components, then its current distribution must be a superposition of the current distributions of said components:

Fig. 3.2 Voltage-domain
characteristics of an example
n-terminal device. With (**a**) its
current distribution, (**b**) the
corresponding cumulative
current distribution, and (**c**)
the corresponding cumulative
power distribution

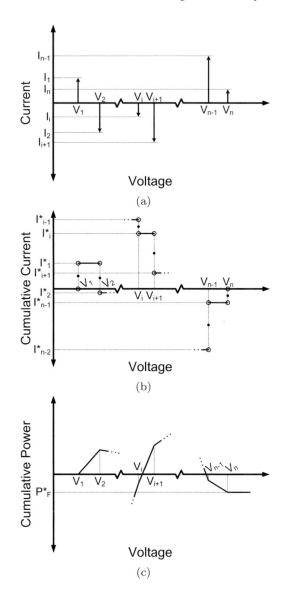

$$I(v) = \sum_{components} I_{component,i}(v),\tag{3.8}$$

with $I_{component,i}(v)$ the current distribution of component i. Naturally, because
integration is a linear operator, the same can be said for the cumulative current-
and cumulative power distributions. This basic property will be used in the next
section to proof a general law of switched-capacitor DC–DC converters.

3.3 Proof of Switched-Capacitor Law

3.3.1 Two-Phase Converter

Figure 3.3 shows an ideal three-terminal step-down DC–DC converter. It has three terminals, with voltages V_{in}, V_{out}, and V_{ss} and respective terminal currents I_{in}, I_{out}, and I_{ss}. This converter's $I(v)$, $I^*(v)$, and $P^*(v)$ are visualized in Fig. 3.4a–c, respectively. From Fig. 3.4c it can be appreciated that, because it is lossless, the converter's cumulative power distribution forms a triangle with the x-axis with a base of $(V_{in} - V_{ss})$ and a height of $(V_{out} - V_{ss})I_{ss}$. The area of this triangle, H_{conv}, can thus be deduced:

$$H_{conv} = \frac{(V_{in} - V_{ss})(V_{out} - V_{ss})(I_{out} - I_{in})}{2}. \tag{3.9}$$

Switched-capacitor DC–DC converters make use of switches and capacitors. The main contribution of the switches is that they generate the correct connections between the capacitors, and the capacitors and the external terminals. In this particular analysis, however, these connections themselves are not relevant. Instead, the importance lies with the voltage at which a connection and subsequent charge transfer take place. Therefore, it is possible to use a simplified flying capacitor element that connects to a certain voltage, rather than another component. In this set-up, connections between flying capacitors simply take place via these voltages.

A lossless two-phase flying capacitor element in steady-state is shown in Fig. 3.5. Here, V_{t1} and V_{t2} are the capacitor's top-plate voltages in phase 1 and 2, respectively, V_{b1} and V_{b2} are similarly the corresponding bottom-plate voltages, and I_C is the capacitor's time-averaged current in either phase. Note that this current is time-averaged relative to a full clock period, not just the phase duration, and that $V_{t2} > V_{t1}$. By convention, the flying capacitor is assumed to be discharging in the first phase, and charging in the second phase. If the opposite is true, I_C will simply be negative. Because this element is losses, the flying capacitor's voltage, V_C, does not change over time, such that $V_C = V_{t1} - V_{b1} = V_{t2} - V_{b2}$.

Figures 3.6a–c illustrate the distributions of this flying capacitor element. It is interesting to point out that in all voltage-domain distributions V_{t1} and V_{b2} can be interchanged without changing the distributions' shape. An ideal two-phase flying

Fig. 3.3 Ideal three-terminal step-down DC–DC converter

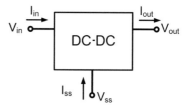

Fig. 3.4 Voltage-domain
characteristics of an ideal
three-terminal step-down
DC–DC converter. (**a**) is its
current distribution, (**b**) its
cumulative current
distribution, and (**c**) is the
corresponding cumulative
power distribution

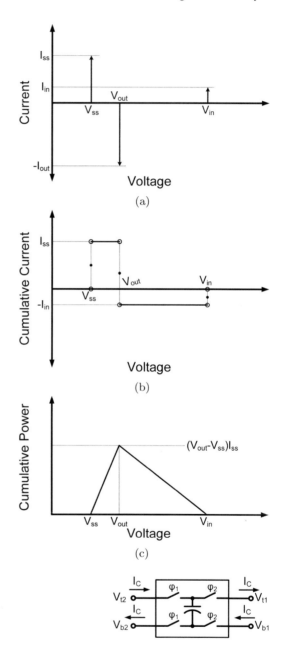

(a)

(b)

(c)

Fig. 3.5 Ideal two-phase
flying capacitor

capacitor consequently has the same time-averaged characteristics to one with its
voltage and BP voltage swing reversed.

From Fig. 3.6c it can be appreciated that a two-phase flying capacitor's cumulative power distribution forms an isosceles trapezoid with the x-axis. The area of this
trapezoid is calculated to be

Fig. 3.6 Voltage-domain
characteristics of an ideal
two-phase, lossless flying
capacitor. With (**a**) its current
distribution, (**b**) its
cumulative current
distribution, and (**c**) its
cumulative power distribution

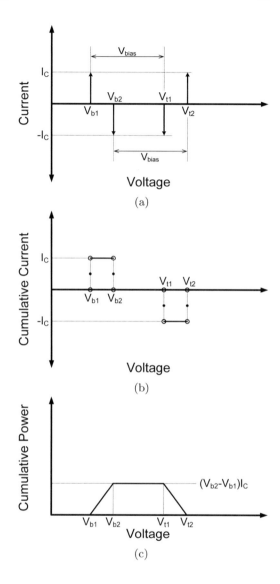

$$H_{cap}^{(2p)} = (V_{b2} - V_{b1})(V_{t1} - V_{b1})I_C. \tag{3.10}$$

Now, because any switch-capacitor topology can be described as a linear combination of these flying capacitor elements, for any such topology H_{conv} must be equal to the sum of its flying capacitors' H_{cap}'s. Combining (3.9) and (3.10), and using the fact that $a_{c,i} = I_{C,i}/I_{out}$, the two-phase law is derived:

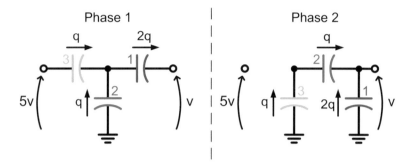

Fig. 3.7 Three-stage DC–DC SC Fibonacci converter

Table 3.1 Capacitor characteristics of three-stage Fibonacci converter

Capacitor	$a_{c,i}$	$v_{bias,i}$	$\Delta v_{b,i}$	Product
1	2/5	1	1	2/5
2	1/5	2	1	2/5
3	1/5	3	2	6/5
Sum				2

$$\sum_{caps,i} v_{C,i} \left(v_{b,i2} - v_{b,i1} \right) a_{c,i} = \frac{1 - iVCR}{2iVCR}, \tag{3.11}$$

where $v_{b,ij} = V_{bj,i}/(V_{out} - V_{ss})$ are the normalized capacitor BP voltages in the jth phase, and V_{ss} is taken as the ground voltage.

To illustrate this, consider the 5:1 Fibonacci converter shown in Fig. 3.7. This converter has a total of three flying capacitors with varying bottom-plate swings, voltages, and charge transferred. An overview of these parameters is given in Table 3.1. As predicted by (3.11), the sum of the product of these properties matches the right-hand side of (3.11). At the same time, the voltage-domain characteristics of the capacitors add up to those of the full converter, as demonstrated in Fig. 3.8.

3.3.2 Multiphase Converter

The relationship of (3.11) can be generalized to any number of phases, by deriving the current distribution of a multiphase flying capacitor element, $I_{mp}(v)$. Consider such an element with p phases, top- and bottom voltages V_{ti}/V_{bi}, and corresponding time-averaged currents in each phase I_{Ci}:

$$I_{mp}(v) = \sum_{phases} I_{Ci} \Big(\delta(v - V_{ti}) - \delta(v - V_{bi}) \Big). \tag{3.12}$$

Fig. 3.8 Voltage-domain characteristics of a three-stage DC–DC SC Fibonacci converter and its capacitors

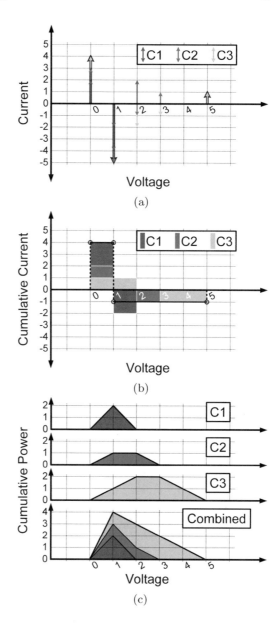

Now, it can be shown that the same distribution can be obtained using $p - 1$ two-phase flying capacitors with the same capacitor voltage, each of which switches between two voltage pairs, V_{tk}/V_{bk} and $V_{t(k+1)}/V_{b(k+1)}$, and with corresponding time-averaged currents, $I_{Ck}^{(2p)}$. Rearranging (3.12) yields

Fig. 3.9 Current distribution
of an ideal p-phase, lossless
flying capacitor. Only
top-plate currents are
portrayed

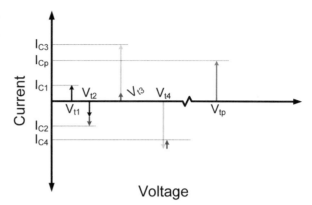

$$I_{mp}(v) = \sum_{k=1}^{p-1} \left[\left(\delta(v - V_{t(k+1)}) - \delta(v - V_{b(k+1)}) \right. \right.$$

$$\left. \left. - \delta(v - V_{tk}) + \delta(v - V_{bk}) \right) \sum_{j=1}^{k} -I_{Cj} \right]. \tag{3.13}$$

Thus, if $I_{Ck}^{(2p)} = \sum_{j=1}^{k} -I_{Cj}$, then it can be appreciated that the right-hand side of (3.13) indeed matches a superposition of $p - 1$ two-phase flying capacitors. This principle is also illustrated in Fig. 3.9. Combining this with (3.10), the $P^*(v)$-area of a multiphase flying capacitor is deduced:

$$H_{cap} = V_C \sum_{k=1}^{p-1} (V_{t(k+1)} - V_{tk}) I_{Ck}^{(2p)}$$

$$= V_C \left(V_{tp} I_{C(p-1)}^{(2p)} - V_{t1} I_{C1}^{(2p)} + \sum_{k=2}^{p-1} V_{tk} (I_{C(k-1)}^{(2p)} - I_{Ck}^{(2p)}) \right)$$

$$= V_C \sum_{phases} V_{tk} I_{Ck}$$

$$H_{cap} = V_C \sum_{phases} V_{bk} I_{Ck}, \tag{3.14}$$

and in turn the general law of switched-capacitor DC–DC converters is obtained:

$$\sum_{caps,i} \left(v_{C,i} \sum_{phases,j} v_{b,ij}\, a_{c,ij} \right) = \frac{1 - iVCR}{2iVCR}. \tag{3.15}$$

Even though the derivation of (3.15) was done for step-down converters here, it is valid for any type of conventional SC DC–DC converter, including step-up converters and negative voltage generators, as long as $I_{C,ij}$ is considered positive when flowing into the capacitor's TP, and I_{out} is positive when flowing out of the converter. In other words, I_{out} should be negative for voltage generators.

3.4 Implications

3.4.1 Bottom-Plate Wolaver Limit

As discussed in Sect. 3.3 there is a distinct symmetry between a two-phase flying capacitors' BP voltage swing and its capacitor's voltage: two two-phase flying capacitors with opposite BP swing and capacitor voltage have identical time-averaged voltage-domain distributions. As a result, it is possible to derive the symmetrical twin of a given two-phase SC topology by interchanging these parameters for every capacitor. The Dickson- and Series-Parallel converter topologies are an example of such twins. That being said, it is possible that this operation does require the addition of buffer capacitors. This is simply due to the fact that the phase domain is averaged-out in the voltage-domain analysis, which means there is no guarantee that the charge transfers of the capacitors match up with each other in the right phase.

Furthermore, this symmetry paves the way to expand the Wolaver relationships for two-phase SC's BP voltage swings. After all, if a two-phase flying capacitor's voltage and bottom-plate swing can be interchanged, any constraint on the capacitor's bias voltages should also be true for its BP voltage swings. Using (3.1) it is consequently conjectured that:

$$\sum_{caps,i} |v_{C,i}| \geq \frac{1 - iVCR}{iVCR} \implies \sum_{caps,i} |\Delta v_i| \geq \frac{1 - iVCR}{iVCR}. \tag{3.16}$$

3.4.2 Topological Efficiency

From Fig. 3.10 it can be appreciated that a capacitor's power distribution area might be opposite in sign compared with the converter's power distribution area. An example of this would be a flying capacitor that is charged at a low BP voltage and discharged at a high voltage within a step-down converter. If this occurs, the capacitor's contribution is effectively counter-productive to the power conversion effort. Needless to say, any counter-productive contribution must be overcome by the other capacitors. A topology with such a capacitor can consequently be considered *topologically inefficient*. In general, topologically inefficient topologies

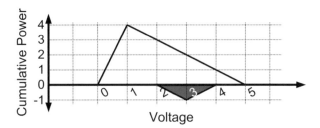

Fig. 3.10 Flying capacitor area with opposite sign relative to converter

will lead to worse performances, although this does not mean there might not be very good practical reasons to opt for them.

Similarly, a capacitor can also have a zero-area cumulative power distribution. While this is preferable over a counter-productive contribution, this capacitor will still take up passive volume, and might even have losses associated with it. The Ladder converter topology has several buffer capacitors that match this description.

With this in mind, a topology is defined to be *topologically efficient* if the following conditions are met:

$$\sum_{caps,i} \left| v_{C,i} \sum_{phases,j} v_{b,ij}\, a_{c,ij} \right| = \left| \frac{1 - iVCR}{2iVCR} \right|, \tag{3.17}$$

$$\forall i \in caps: \quad v_{C,i} \sum_{phases,j} v_{b,ij}\, a_{c,ij} \neq 0. \tag{3.18}$$

Comparing (3.17)–(2.59), it can also be appreciated that for two-phase converters, all topologically efficient topologies have identical performance in the low power-density regime assuming constant energy-density capacitors. As such, it should come as no surprise that the Series-Parallel-, the Fibonacci-, and the Dickson converter, which were analyzed in this regime in Sect. 2.3.3 are all topologically efficient families.

Considering (2.58), however, it is clear that the same cannot be said for converters using more than two phases:

$$\sum_{phases,j} v_{b,ij}\, a_{c,ij} \not\equiv \sqrt{\sum_{phases,j} a_{c,ij}^2 \sum_{phases,j} \Delta v_{ij}^2} \tag{3.19}$$

This reveals an opportunity for multiphase converters to perform better than the apparent limit, which will be exploited in Chap. 4.

3.4.3 Topology Synthesis

The proposed voltage-domain analysis can also be used as a visual tool for the synthesis of switched-capacitor converter topologies. For a desired iVCR, one simply has to draw the corresponding triangle of the cumulative power distribution and come up with a set of isosceles trapezoids that add up to this triangle. Of course, the VDA does not provide a strategy for arriving at such a set, but this is best left to a human's puzzle-solving mind. The easiest way to do this in practice is by drawing the triangle on a grid with a sufficiently high resolution, and then filling the triangle up, similar to Fig. 3.8c. In addition, it can be useful to also draw the trapezoids on a separate plot.

When a set of isosceles trapezoids is found, each trapezoid can be translated to at least one, often two, distinct two-phase flying capacitors. The same combination of trapezoids can thus lead to multiple topologies. However, as was discussed Sect. 3.4.1, there is no guarantee that the set of two-phase flying capacitors is compatible in the phase domain. As a result, some buffer capacitors might still be required.

If some two-phase flying capacitors have the same capacitor voltage, they can be merged into a multiphase flying capacitor, thereby opening the opportunity for multiphase topology synthesis.

3.4.4 Fundamental Trade-Offs

Most importantly, (3.11) and (3.15) expose a fundamental trade-off between a topology's capacitors' voltages, BP swings, and its charge multiplier vector. For example, for low-power monolithic converters using a single-capacitor implementation, emphasis is usually put on minimizing the BP voltage swings and charge multiplier elements in accordance with (2.44). The introduced switched-capacitor law tells us that this will inevitably lead towards topologies with larger capacitor voltages, like the Dickson converter [MSBS14]. Similarly, as shown in Sect. 2.3.3, the Series-Parallel has excellent performance at high power-densities, precisely because its larger BP swings are not an issue in this regime.

3.5 Conclusions

This chapter introduced a new type of analysis method for DC–DC power converters, called a voltage-domain analysis, where the time-averaged current distributions of systems and their elements are compared with each other in the voltage domain. Thanks to this analysis, a general law of conventional switched-capacitor DC–DC converter topologies was proven that relates the flying capacitors' voltages, BP swings, and charge multiplier elements with the topology's ideal voltage conversion ratio. The analysis also revealed a symmetry between the capacitor's voltage and BP swing in two-phase SC converters that was subsequently used to conjecture an expansion of the Wolaver limit of capacitor voltages to BP swings.

The general law was in turn used to come up with a definition of what it means to be efficient from a purely topological point of view. Surprisingly, when combined with the results of Chap. 2, it was found that all topologically efficient two-phase converters have identical performance in the low power-density regime under certain assumptions, and that multiphase converters subsequently have an inherent advantage. In addition, the general law provided some additional insights to why some topology families are favored in certain situations over others. Finally, the introduced voltage-domain analysis was briefly discussed as a visual tool for the synthesis of switched-capacitor converter topologies.

Chapter 4
Scalable Parasitic Charge Redistribution

In this chapter a first advanced multiphasing technique, called Scalable Parasitic Charge Redistribution (SPCR), is introduced, that reduces the parasitic coupling losses in fully integrated switched-capacitor DC–DC converters up to any desired level. This is accomplished by continuously recycling parasitic charge in-between phase-shifted converter cores over multiple phases. Because this subsequently transforms the switched-capacitor converter into a multiphase converter, it is no longer bound by the limits described in Chap. 2, and can consequently achieve efficiencies previously deemed impossible.

The model of Chap. 2 is updated to include transistor leakage, which is found to be a limiting factor to how much charge can efficiently be recycled. Using this updated model, the effectiveness of SPCR is demonstrated over a large range of technology parameters. In addition, the implementation of SPCR can be done with little area overhead thanks to the use of charge redistribution buses. A 2:1 converter is fabricated in a 40 nm bulk CMOS technology that verifies SPCR by achieving a record efficiency for fully integrated closed-loop SC converters of 94.6%.

4.1 Motivation

As discussed in Chap. 1, the delivery of power to integrated circuits has become increasingly difficult for a wide range of applications. At the core of this issue lies the continued scaling of supply voltages. Lowering the supply voltage of applications has simultaneously lead to an increase of their intake current, which has in turn increased losses related to the power delivery network. As a solution to this problem, many have suggested moving part of the voltage conversion on chip using monolithic PMUs [Car14, CFM+10, SVBM+11]. Doing so would reduce the intake current by the achieved on chip Voltage Conversion Ratio (VCR) and

© Springer Nature Switzerland AG 2020
N. Butzen, M. Steyaert, *Advanced Multiphasing Switched-Capacitor DC-DC Converters*, https://doi.org/10.1007/978-3-030-38735-8_4

allow for closer regulation. PDN and regulator induced voltage margins can thus be significantly reduced.

At the same time, there has been a rising interest in energy scavenging for internet-of-things (IoT) applications or wireless sensor nodes. These nodes are preferably made as small as possible but also need to last for multiple years. Because the battery that would be required for this lifetime is often impractical, using the energy around the node itself presents itself as a natural solution. Combined with a fully integrated PMU, the total size could be shrunk to cubic millimeter size [FKC⁺13].

In both cases, though, it is crucial that the efficiency of the PMU is sufficiently large, especially considering the lower quality of passives and larger parasitics in the monolithic context. Consequently, the PMU efficiency is often considered to be its most important specification [Car14].

For switched-capacitor converters, Chap. 2 established the presence of an efficiency ceiling in the low power-density regime. According to (2.33), this maximum obtainable efficiency of a SC DC–DC converter is described by

$$\eta_{max} = \frac{1}{1 + 2\sqrt{\alpha_{par} K_{par} K_C}}. \tag{4.1}$$

Thus, the larger α_{par}, the lower the efficiency that can be obtained. With α_{par} typically around 1.5% for Metal-Oxide-Metal (MOM) and Metal-Insulator-Metal (MIM), and 7% for Metal-Oxide-Semiconductor (MOS) capacitors, a 2:1 SC converter could theoretically achieve efficiencies up to 89% and 79%, respectively. Due to additional losses (control, leakage, interconnect, ...) the effective efficiency ceiling is lower, but its existence is still confirmed by previous work [STM⁺15] [SBM⁺]. The highest reported fully integrated closed-loop converter efficiencies in baseline CMOS are 87% [VBS10a], although at a more favorable VCR of 3:2, and 85% [SM14a], both using MIM capacitors. Higher efficiencies have been demonstrated using either open-loop converters with VCR's very close to 1:1 (95% in 16:15) [JSB16], or high-density Deep-Trench (88% in 2:1) [AKK⁺17] or Ferro-Electric (91% in 2:1) [EDBC13] capacitors, which have reportedly up to 25 times lower α_{par}. However, these capacitors are not part of baseline CMOS and thus require additional masks and costs.

Due to its importance on the achievable efficiency, some have suggested to short the BP nodes of two converters in anti-phase during the dead time in-between phase transitions, effectively redistributing half of the parasitic charge from the discharging BP capacitor to the charging one [AKK⁺13, See09]. The supply voltage consequently only needs to supply the remaining half to charge the BP capacitor, resulting in a 2× reduction in parasitic coupling losses. While effective, this method does not scale to higher levels of redistribution because it still uses the same two-phase control signals for the converter.

This chapter is organized as follows. In Sect. 4.2, a technique that reduces the parasitic coupling losses up to any desired level, called Scalable Parasitic Charge Redistribution (SPCR), is introduced. Its effect on the design space of fully integrated SC converters will be explored in Sect. 4.3. Section 4.4 goes into more detail how SPCR can be efficiently implemented using an example implementation. Measurement results, verifying the effectiveness of SPCR in a realized converter, are discussed in Sect. 4.5. Finally, Sect. 4.6 highlights the important conclusions.

4.2 Working Principle

4.2.1 Regular Switched-Capacitor Converters

In Fig. 4.1a the general working principle of a regular SC converter's flying capacitor is portrayed. Each phase, a certain amount of charge is pumped by the flying capacitor, depending on the voltage mismatch between both domains and the size of the capacitor, C_{fly}. However, in order to keep this process running, once every other phase, charge needs to be invested into the parasitic BP capacitor. This parasitic charge, q_{par}, scales with the size of the parasitic coupling, C_{par}, and the voltage step between both domains, ΔV. The former is largely determined by the process technology and the type of capacitors used. Instead, ΔV will be key in reducing the associated BP losses up to any desired level. Figure 4.1b shows an alternative representation of the same converter but with additional emphasis on the bottom-plate voltage, V_b. This representation will be used throughout this chapter.

Figure 4.2 shows a multiphased, or time-interleaved, converter. Here, the converter is split up into several smaller converter cores. Each clock edge, the two cores that have been in the high/low state the longest, transitions to the next state by fully charging/discharging their V_b to V_{high}/V_{low}. Consequently, while this technique does reduce the output voltage ripple significantly [VBS09], the parasitic coupling losses remain unchanged.

4.2.2 Charge Redistribution

The general idea of SPCR is straightforward: Instead of having multiple time-interleaved converter cores working in parallel to one another, the cores will actively and continuously redistribute their parasitic charge among each other to reduce the parasitic coupling losses and thus enhance their efficiency. To this end, a dedicated BP charging and a dedicated BP discharging state are introduced (Fig. 4.3). Rather than transitioning from a high state directly to a low state, a core will first enter the dedicated BP discharging state. Likewise, a core will go through the BP charging state when going from the low to the high state.

In Fig. 4.4 an example SC converter using SPCR is shown. Cores that are neither in the regular high, nor in the regular low state, are instead in the BP charging or discharging state. Here, all the regular power transistors are non-conducting and

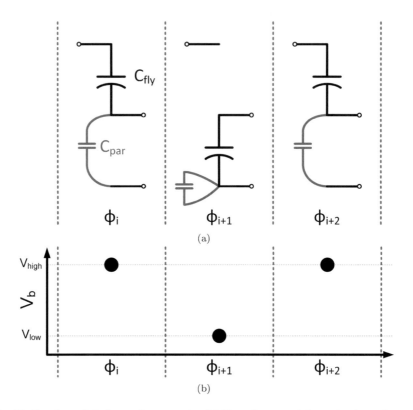

Fig. 4.1 Regular switched-capacitor converter. (**a**) Normal representation using lumped capacitors, and (**b**) alternative representation used in this chapter

the core itself can only be at a set number of intermediate levels, chosen during design time. At every clock edge, each BP charging core is paired up with the BP discharging core which is in the closest, yet higher intermediate level. By shorting the BP nodes of each pair, their V_b's average out by transferring charge from the BP discharging to the BP charging core. This is called a charge redistribution step (CRS) and results in all paired BP charging cores going up, and all paired BP discharging cores going down one intermediate level. BP charging/discharging cores which are already at the highest/lowest intermediate level, and can consequently pair up no more, are instead pulled up/down to the high/low state. Furthermore, to keep this process going, every two phases, the two cores that have been in the high/low state for the longest time, are transferred to the BP discharging/charging state. The end result is that the low to high transition is now completed approximately adiabatically using a fixed number of CRS, equal to the number of intermediate levels and that V_{high} only needs to supply enough charge to pull the core up to the high state, which is in general $(CRS + 1)$ times lower than the charge without SPCR. The parasitic coupling losses are consequently also reduced by the same factor:

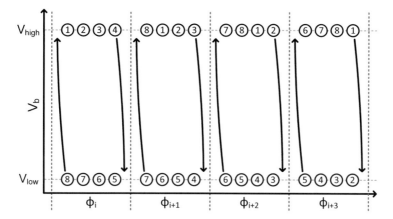

Fig. 4.2 An example multiphased switched-capacitor converter with eight cores. Each labeled circle represents a different converter core. Arrows represent actions during phase transition

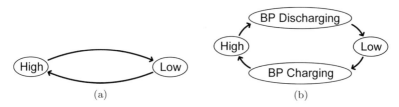

Fig. 4.3 State transitions of (**a**) a regular SC converter or a time-interleaved converter core and (**b**) an SPCR converter core

$$P_{par,SPCR} = \frac{P_{par,regular}}{CRS + 1}. \tag{4.2}$$

Because all cores are phase-shifted versions of each other, the necessary connections between cores and timing of the charge exchanges are known at design time, significantly simplifying the design of this kind of SC converter. The phases as shown in Fig. 4.4 are also stable and require no initialization. This can intuitively be explained as follows: Because every CRS is an averaging operation, the voltage of an intermediate level will be the average of its surrounding levels. The intermediate levels will subsequently naturally spread evenly between the boundary conditions of V_{high} and V_{low}.

SPCR can be implemented no matter the switched-capacitor topology, and thus VCR, by applying the technique to each unique capacitor separately. When multiple unique capacitors have the same bottom-plate swing, as is the case with the Dickson topology, they can even use SPCR with each other. For a SC converter using SPCR there are two important design parameters: The total number of cores, N, and the number of intermediate levels or charging steps, CRS. The only condition for SPCR

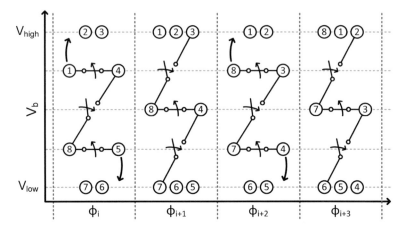

Fig. 4.4 An example switched-capacitor converter using Scalable Parasitic Charge Redistribution with eight cores and three charge redistribution steps. Arrows represent actions during phase transitions

to work in theory is given by Eq. (4.3). If it is not met, there are simply not enough cores to pair up each phase transition.

$$N \geq CRS + 1. \tag{4.3}$$

4.2.3   Top Plate

While in the literature the losses corresponding to the parasitic coupling of the flying capacitor are often referred to as bottom-plate losses, the equivalent parasitic capacitor is not necessarily entirely or even partly connected to the flying capacitor bottom-plate node. Metal-finger based MOM capacitors, for example, generally have equal coupling on the top- and the bottom-plate. At the same time, capacitors with an asymmetrical parasitic coupling can still be connected such that the node with the highest voltage has most of the coupling. Using SPCR with capacitors which have significant TP coupling raises the interesting question: Should the charge redistribution (also) take place at the top-side of the flying capacitor?

In Fig. 4.5 a single CRS of a core in the BP charging state which has a certain TP coupling C_{top} is shown. As demonstrated in the same figure, this situation can simply be modeled with an equivalent $C_{bot,eq}$, that is the series of its parts. If C_{top} is sufficiently small compared to C_{fly}, $C_{bot,eq}$ will be approximately equal to C_{top} and the voltage step on the top plate can be approximated by the voltage step on the BP plate. The parasitic charge on the TP can thus indeed be redistributed through the BP node. However, two important second-order effects occur when doing so. First, for step-down converters, the total TP swing is generally larger than the total BP swing due to the fact that such a converter transfers charge from a high- to a low voltage

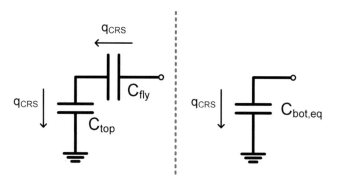

Fig. 4.5 Charge transfers during single charge redistribution step of a flying capacitor with parasitic top-plate coupling and the equivalent bottom-plate coupling

by charging its flying capacitors in a high- and discharging the same capacitors in a low voltage domain. Consequently, this pumped charge, ultimately meant for the converter output, is added as an extra voltage difference to the TP swing, as shown in Fig. 4.6a. This extra swing will not be redistributed through this method which means that more charge will need to be supplied in the transition to the high state. The parasitic coupling losses will thus be higher than described by Eq. (4.2). At the same time, though, a CRS of the TP coupling through the BP node also charges C_{fly} (Fig. 4.6b). The authors refer to this effect as *parasitic charging*. In the case of a step-down converter this additional charge is generally in the same direction as the regular pumped charge (occurring in the high and low state) and thus more charge is pumped to the converter output than usual. In terms of losses, both effects were found to cancel each other out. In fact, in a direct comparison of a converter with only TP coupling to one with only BP coupling, no significant change in efficiency was witnessed. The TP coupling converter does, however, always produce a higher output voltage.

Redistributing charge from the TP and BP nodes simultaneously has the undesirable effect of also redistributing charge on C_{fly} meant for the output of the converter. This significantly increases the converter's output resistance and should therefore not be considered.

4.3 Switched-Capacitor Model and Optimization with SPCR

Combining (2.32) with (4.2), allows one to deduce the minimum normalized losses under the model described in Chap. 2:

$$P_{N,min} = 2\sqrt{\frac{\alpha_{par} K_{par} K_C}{CRS + 1}}. \tag{4.4}$$

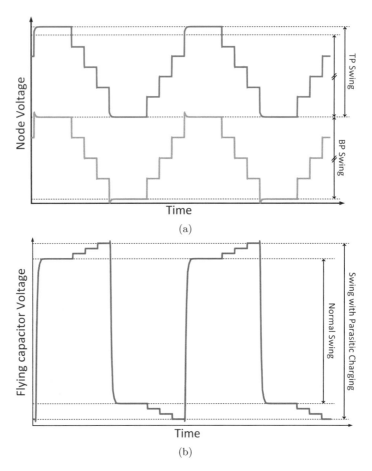

Fig. 4.6 Simulated waveforms of (**a**) the top- and bottom-plate node voltages and (**b**) the flying capacitor voltage, of a SPCR SC converter redistributing TP charge through the BP node

In other words, by increasing the number of charge redistribution steps, the normalized losses could be set to any desired level. Needless to say, this apparent lack of trade-off seems highly implausible and hints to the need to expand the common switched-capacitor model to other loss contributors. For SPCR, a practical trade-off is the addition of extra switches and thus switch area. Therefore the importance of transistor leakage is investigated in this section.

4.3.1 Transistor Leakage

A transistor in the non-conducting state has some leakage current through its channel, I_{subt}, and a gate tunneling current to the gate, I_{rev}. In the conductive state there also is a gate tunneling current, I_{gate}, from the gate to the channel [RMMM03]. The associated losses are given by

$$P_{l,nc} = \sum_{switches,i} (1 - D_{cond,i}) V_{R,i} \left(I_{subt,i} + I_{rev,i} \right) \tag{4.5}$$

$$P_{l,c} = \sum_{switches,i} D_{cond,i} V_{trans,i} I_{gate,i} \tag{4.6}$$

with $D_{cond,i}$ the switch duty cycle for which it is conducting, and $V_{trans,i}$ the transistor driving voltage. Assuming that V_{trans}, V_R, and D_{cond} are identical for all switches, the total leakage losses can be simplified to

$$P_{leak} = \alpha_l(D_{cond}) G_{tot} V_{out}^2, \tag{4.7}$$

where α_l is the normalized average leak-to-conductance conductance ratio,

$$\alpha_l(D) = (1 - D)\alpha_{l,nc} \frac{V_R^2}{V_{out}^2} + D\alpha_{l,c} \frac{V_{trans}^2}{V_{out}^2}, \tag{4.8}$$

and $\alpha_{l,c}$ and $\alpha_{l,nc}$ are the conductance ratios of the conducting- and non-conducting phase, respectively. Similarly to parasitic coupling- and transistor driving losses defined in Sect. 2.2.2, these leakage losses are considered extrinsic to the output impedance model.

4.3.2 Regular SC Optimization

The optimization of Sect. 2.3 is revisited with the addition of transistor leakage losses. This yields the following set of equations:

$$\frac{\partial P_N}{\partial f_{sw}} = 0 \Leftrightarrow P_{SSL} = P_{par} + P_{trans}, \tag{4.9}$$

$$\frac{\partial P_N}{\partial G'_{tot}} = 0 \Leftrightarrow P_{FSL} = P_{leak} + P_{trans}, \tag{4.10}$$

$$\frac{\partial P_N}{\partial P_D} = 0 \Leftrightarrow P_{FSL} + P_{SSL} = P_{par} + P_{trans} + P_{leak}. \tag{4.11}$$

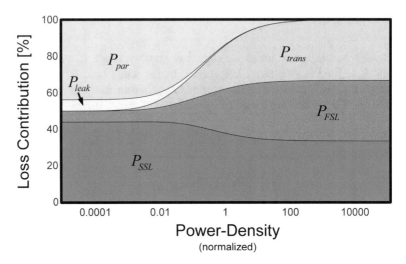

Fig. 4.7 Scaling of the relative loss contributions of an optimized fully integrated switched-capacitor converter versus power-density, which includes transistor leakage losses

Similar to the optimization without leakage, it can appreciated that the total conductance losses should still always be at least as large as the sum of the others combined, no matter the power-density. Likewise, the maximum efficiency is achieved when the output power-density tends to zero together with the dynamic transistor charging losses, and f_{sw} and G'_{tot} will be such that $P_{SSL} = P_{par}$ and $P_{FSL} = P_{leak}$, respectively. The corresponding minimal normalized losses are

$$P_{N,min} = \frac{2\sqrt{P_{SSL}P_{par}} + 2\sqrt{P_{FSL}P_{leak}}}{P_{out}} \tag{4.12}$$

$$P_{N,min} = 2\sqrt{\alpha_{par}K_{par}K_C} + 2\sqrt{\alpha_l(D_{cond})K_S}. \tag{4.13}$$

Thus, similar to the capacitors, the quality of the transistors, determined by the conductance ratio, puts a limit on the maximum obtainable efficiency of a switched-capacitor converter. Figure 4.7 also illustrates this by showing the relative scaling of the losses versus power-density for an example design. Moreover, compared with Fig. 2.6b, it is easy to see that the high power-density regime has remained unchanged.

4.3.3   SPCR Optimization

Using Scalable Parasitic Charge Redistribution changes the losses of an SC converter in three ways. First, the parasitic coupling losses are reduced according to (4.2), assuming no second-order effects, discussed in Sect. 4.2.3, take place. In addition, due to the dedicated BP charging/discharging state, the conduction duty cycle of the regular power transistors in a two-phase converter is lowered to

$$D_{cond} = \frac{N - CRS}{2N},$$
(4.14)

which will increase P_{FSL} and change α_l. Finally, there is a need for charge redistribution transistors (CRTs). Because each of these transistors is shared between two cores, only $CRS/2$ CRTs per core are needed to perform the parasitic charge exchanges. $N \cdot CRS/2$ extra transistors are consequently required in total, which will all add transistor driving- and leakage losses. Assuming each CRT is sized such that 3 RC time constants equal one CRS time period, the total CRT conductance, G_{CRT}, and associated losses, using (2.24) and (4.7), can be respectively described by

$$G_{CRT} = 3\alpha_{par} f_{sw} C_{fly} \frac{N \cdot CRS}{2},$$
(4.15)

$$P_{trans,CRT} = 3\alpha_{par} f_{sw}^2 C_{fly} E_{on} R_{on} \frac{N \cdot CRS}{2},$$
(4.16)

$$P_{leak,CRT} = 3\alpha_{par} \alpha_l (N^{-1}) f_{sw} C_{fly} V_{out}^2 \frac{N \cdot CRS}{2}.$$
(4.17)

The scaling of each loss contributor that is influenced by either N or CRS is summarized in Table 4.1. No straightforward model is found to determine the optimal N or CRS, which means that in a practical design a numerical optimizer is required. That being said, the optimal CRS and N definitely go down with increasing power-densities. This can explained by the fact that parasitic coupling losses are increasingly less important, while $P_{trans,CRT}$ scales with f_{sw}^2.

For the optimal normalized losses of two-phase converters, the following approximation, derived using asymptotic analysis, is proposed:

Table 4.1 Summary of loss contributors influenced by SPCR design parameters

Loss contributor	Scaling
Parasitic bottom-plate	$\dfrac{1}{CRS + 1}$
Power transistor FSL	$\dfrac{2N}{N - CRS}$
Power transistor leakage	$\alpha_l \left(\dfrac{N - CRS}{2N} \right)$
CRT overhead gate	$N \cdot CRS$
CRT overhead leakage	$\alpha_l \left(\dfrac{1}{N} \right) \cdot N \cdot CRS$

$$P_{N,SPCR,min} \approx \frac{3\sqrt{\alpha_{par}K_{par}K_C}}{\sqrt[6]{\dfrac{K_{par}}{\alpha_l(0)}}} + 2\sqrt{\alpha_l(2^{-1})K_S}. \qquad (4.18)$$

Note the similarities between (4.13) and (4.18). Because the duty cycle of a two-phase converter is 0.5, the second term is the same in both instances, while the first is $\frac{2}{3}\sqrt[6]{\frac{K_{BP}}{\alpha_l(0)}}$ times smaller in the SPCR case. Figure 4.8 provides a visual comparison. As expected, for very large conduction ratios, the second term in (4.18) dominates, and the losses are very similar. For smaller ratios, however, SPCR significantly reduces the impact of the parasitic coupling, and allows for much higher efficiencies to be achieved. In general, the higher the transistor quality, the higher the gain with SPCR.

In Fig. 4.9, a comparison is made for practical power-densities. SPCR increases the efficiency over the entire power-density range of $10\,\mu\text{W/mm}^2$ to $10\,\text{W/mm}^2$ and for all parameter variations. At high power-densities, the improvement due to SPCR is invariant to the conductance ratio, which means low-V_T transistor devices can be used. Furthermore, the impact of α_{par} is noticeably reduced. For lower power-densities, on the other hand, the use of low-power or high-V_T devices provides a significant efficiency boost.

4.4 SPCR Implementation

To verify the obtainable efficiencies of the SPCR technique, a fully integrated 2:1 SC converter was designed using 16 cores and 9 CRS, thus reducing the parasitic coupling losses tenfold. A system overview of the converter is shown in Fig. 4.10. Note that because SPCR naturally extends on the time-interleaving concept, no decoupling is used at the output of the converter.

4.4.1 Charge Redistribution Bus

For the charge exchanges to take place, a total of 72 different core interconnections are needed, which will all add overhead area and extra losses due to charging and discharging of their parasitic coupling. Instead, 8 charge redistribution buses (CRBs) are used, as shown in Fig. 4.11. The principle of a CRB is very similar to a data bus: When two BP nodes need to be shorted, they are both connected to the same bus. Furthermore the bus they use depends on their resulting intermediate voltage level after their V_b's average out. The end result is that significantly less area overhead is needed and that the voltage swing, and associated parasitic loss, on each CRB is approximately zero, effectively making them DC voltage rails. Normally this

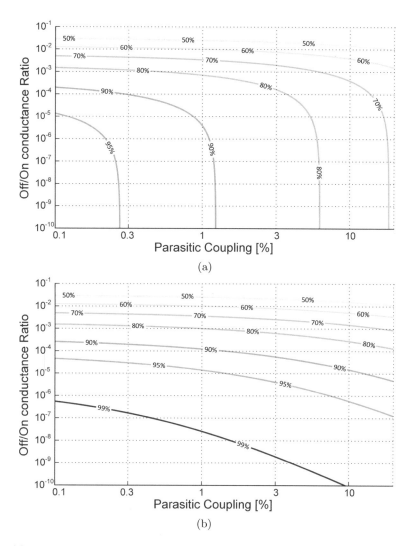

Fig. 4.8 Maximum achievable efficiency of (**a**) a regular 2:1 SC converter and (**b**) a 2:1 converter using SPCR

would require 9 buses (one for each intermediate level). In this design, however, the $0.5xV_{out}$ bus is replaced with a short connection between each in- and anti-phase pair which are already placed physically close to each other because they share most control signals. Also, note that while with this topology only one set of CRBs is necessary, if a topology has more than one capacitor with a unique BP swing, each will require its own set of CRBs.

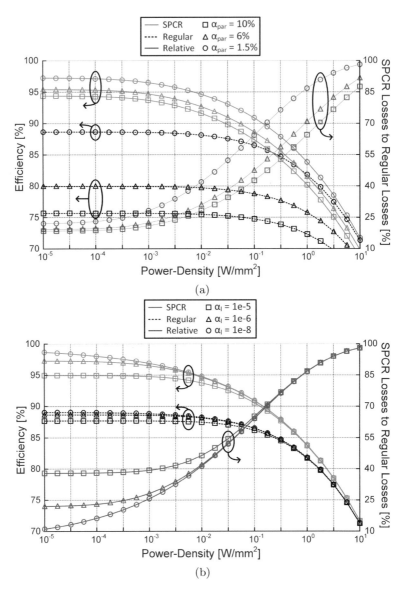

Fig. 4.9 Comparison of the efficiency and normalized losses of a regular 2:1 SC converter to one with SPCR for (**a**) a fixed α_l of 1e−6 and (**b**) a fixed α_{par} of 1.5%

4.4.2   Control

The output of the converter is regulated at a fixed voltage V_{ref} using a lower-bound hysteretic controller clocked at 50 MHz [MD99]. Because for an uneven CRS a

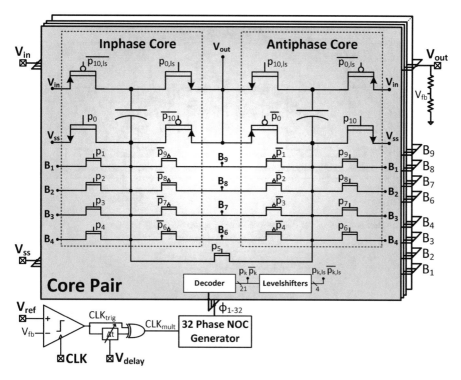

Fig. 4.10 System overview of the 2:1 converter, showing the controller and transistor-level implementation of the converter cores

Fig. 4.11 Core interconnect schema using charge redistribution buses and a direct 0.5xV_{out} connection between in- and anti-phase cores

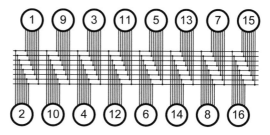

core is only switched to V_{out} once every other phase, each comparator trigger event needs to cause two-phase transitions to assure a fast response time. This is done using a XOR-based edge detector with a variable delay block, which passes two pulses at a time to a 32-phase non-overlapping-clock (NOC) generator. The NOC generator, implemented by 32 ring-connected dynamic master-slave flip-flops (FFs) shown in Fig. 4.12, is designed such that the width of the pulse determines the non-overlapping time between each of the one-hot coded 32 phases (Fig. 4.13a). Note that in order to function properly, the FFs need to be initialized such that all but one are in the same state.

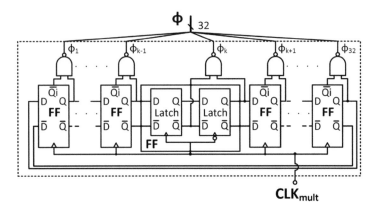

Fig. 4.12 Implementation of the 32-phase non-overlapping-clock generator

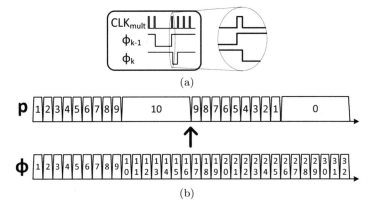

Fig. 4.13 Details of (**a**) the non-overlapping phase generation by pulse-width modulation of the clock, and (**b**) a local decoder schema

All 32-phase signals are subsequently used to locally generate the correct control signals according to the decoding schema shown in Fig. 4.13b or a phase-shifted version thereof. The decoder itself can be efficiently implemented using a total of 9 NAND2 and 9 INV gates for the CRT signals, and 2 Set-Reset (SR) latches for the regular power transistors. Two capacitively coupled levelshifters shift the necessary signals to the high domain (V_{out} to V_{in}) [MPK$^+$15]. While efficient, this type of levelshifter does tend to limit the converter's input range. This can be explained as follows: As ($V_{in} - V_{out}$) gets larger relative to V_{out}, the coupling capacitors will achieve less and less of the full swing in the higher domain. Furthermore, thanks to their increasing V_{gst}, the top cross-coupled inverter pair's gain will also increase relative to the gain of the bottom driving buffers. Combined with differences in arrival times of the phase and anti-phase component of the input of the levelshifter, the signal will eventually no longer be passed on to the higher domain properly.

Table 4.2 Transistor width
sizes per core in μm

	PMOS	NMOS
Power transistors	57.6	32.4
CRT at B_6/B_4	16.38	5
CRT at B_7/B_3	5.92	2.04
CRT at B_8/B_2	3.06	1.33
CRT at B_9/B_1	2.28	1.15
CRT at B_5		2

4.5 Experimental Verification

The design is realized in a 40 nm baseline CMOS process, using 10 nF of MOM capacitance. An overview of the transistor widths is given in Table 4.2. It can be appreciated that, because the CRTs are driven using V_{out} and V_{ss}, they are progressively sized larger as the corresponding bus voltage nears the halfway point. All transistors are realized using high-V_T devices to lower the parasitic conductance ratio, with the only exception being the CRT that connects the bottom-plate nodes of a core pair together, where instead a low-V_T transistor is used to improve the V_{gst}. In total, the CRTs add up to less than 18% of the total transistor width.

Using (4.13), the theoretical maximum efficiency of the technology was determined to be approximately 89% for a 2:1 conversion. The converter, shown in Fig. 4.14, measures 0.94×2.59 mm^2 without bond pads and has a total active area of 2.2 mm^2. The core pairs are distributed in two rows with the CRBs and control signal interconnections in-between. Their relative placement was optimized for minimal signal interconnect length.

4.5.1 Measurement Setup

An overview of the measurement setup is given in Fig. 4.15. A Keithley 2450 source meter provides the input voltage and measures the input current [Kei11]. The clock reference is generated using a Rigol DG3101A arbitrary waveform generator [RIG08], while the reference- and delay voltage are set using two trim resistors powered by an Agilent E3610A DC power supply [Key07]. To accurately measure the efficiency, a system of on-chip Kelvin contact is used, which are measured using three Keysight 34401A digital multimeters [Key12]. The load is a PCB-level resistor array that also includes trim resistors. Another digital multimeter of the same make, measures the load current.

For the load transient measurements, a PCB-level MOSFET, selected for low on-resistance, was put in series with the resistor array. A second waveform generator was then used to trigger the said MOSFET periodically at a frequency much lower than the converter control frequency, while the output voltage waveform was captured with a Keysight MSO-X 4101A oscilloscope [Key17], and an Agilent

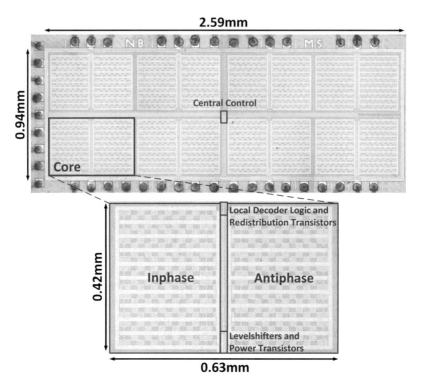

Fig. 4.14 Annotated micrograph of the fully integrated 2:1 SC converter using SPCR, measuring 2.4 mm^2 without bond pads

N2796A active probe [Key18]. During all other measurements, the MOSFET was kept in the linear regime.

4.5.2 Efficiency

Figure 4.16 shows the converter's measured closed-loop efficiency versus output power for an input voltage and output voltage of 1.855 V and 900 mV, respectively. The efficiency includes all system losses. A record efficiency of 94.6% is achieved for output powers of 2.7–3.15 mW. Due to the input-referred quiescent current of 15 μA, the measured efficiency remains high for a wide range of output powers: At 13% of the maximum output power of 3.85 mW, the efficiency is still above 90%.

The converter also maintains a high efficiency over its entire usable input voltage range of 215 mV, as shown in Fig. 4.17. The lowest efficiency, measured at the maximal V_{in} of 2.07 V, was determined to be 88.6%, which is higher than the previous state-of-the-art [SBM$^+$]. SPCR does consequently not solely increase the

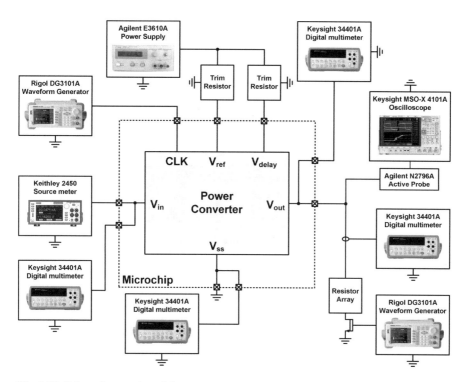

Fig. 4.15 Schematic overview of the measurement setup

efficiency of the nominal design point, but also boosts the efficiency over the entire voltage range. The VCR-to-$iVCR$ ratio, an indication of the converter's conduction losses, reaches a maximum value of 97%, which corresponds to an equivalent drop-out voltage of less than 30 mV. This low value can be partly attributed to the symmetrical parasitic coupling of MOM capacitors and the parasitic charging effect discussed in Sect. 4.2.3. Also, note that at the nominal design point, the conduction losses account for slightly more than 50% of the total losses, which conforms to a properly optimized design as discussed in Sect. 4.3.2.

4.5.3 Controller

To validate the hysteretic controller, it is tested under worst-case load-regulation conditions, as shown in Fig. 4.18. Here, the load current is switched from the nominal load of 4.25 mA to zero and back with a transient time of 8 ns and with a fixed V_{in} of 1.855 V and V_{ref} of 900 mV. Without the use of an internal or external output capacitor, C_{DC}, the droop and overshoot are only 21 and 18 mV, respectively,

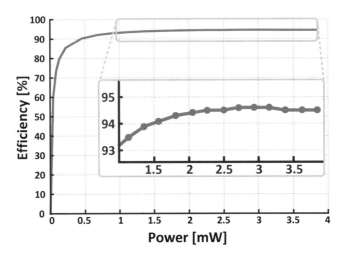

Fig. 4.16 Measured efficiency versus output power for $V_{in} = 1.855$ V and $V_{out} = 900$ mV

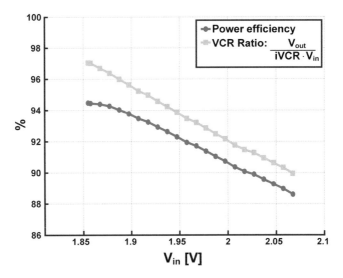

Fig. 4.17 Measured efficiency and VCR-to-$iVCR$ ratio versus V_{in} for a constant load impedance, R_L, and V_{ref} of 212 Ω and 900 mV respectively

showing the fast response capabilities of the lower-bound hysteretic control. The worst-case output voltage ripple, for the nominal input voltage of 1.855 V, was found to be 18 mV.

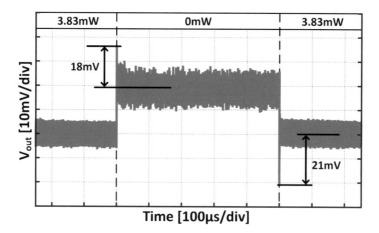

Fig. 4.18 Full load-step transient response with $V_{in} = 1.855$ V and $V_{ref} = 900$ mV

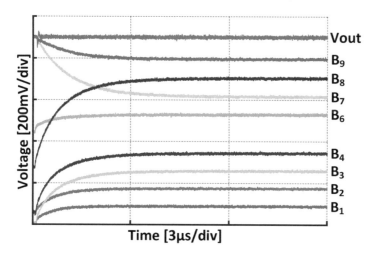

Fig. 4.19 Start-up of the charge redistribution buses with nominal-operation control signals

4.5.4 SPCR Start-Up

As mentioned in Sect. 4.2.2, the use of SPCR does not require any kind of initialization because it is inherently stable due to the averaging operation. To support this claim, the start-up of the charge redistribution buses was measured using regular nominal-operation control signals. As can be seen in Fig. 4.19, the CRBs voltages converge. Furthermore, their steady-state values are evenly spread between ground and V_{out}, which confirms the basic working principle of the CRBs.

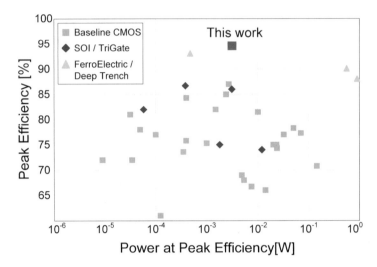

Fig. 4.20 Visual comparison of this work to the state-of-the-art of fully integrated closed-loop switched-capacitor (SC) converters [SBM$^+$]

4.5.5   Comparison

Finally, this work is compared to the state-of-the-art of fully integrated closed-loop SC converters in Fig. 4.20 and Table 4.3 [BS16a, BS16c]. Thanks to the presented advanced multiphasing SPCR technique, the realized converter achieves a higher efficiency than any other fully integrated SC regulator, including those using Deep-Trench and Ferro-Electric capacitors which require extra masks. The in this work achieved efficiency of 94.6% is consequently a new record. When comparing for the same 2:1 ratio, thus negating topological differences (K_{par}, K_C, K_S), the use of SPCR leads to a 68% and 42% reduction in normalized losses compared to bulk CMOS and Ferro-Electric regulators, respectively, which shows the significant advantage SPCR provides.

Furthermore, because SPCR extends naturally on time-interleaving, the presented work realizes a small output voltage ripple without the use of any load capacitance, C_{DC}, and a higher output power and power-density compared to other highly efficient bulk CMOS regulators.

4.6   Conclusion

This chapter briefly discussed the need for highly efficient fully integrated voltage regulators and the parasitic substrate coupling as limitation for high efficiency in SC converters. An advanced multiphasing technique called Scalable Parasitic Charge

Table 4.3 Comparison to state-of-the-art

	This work [BS16a, BS16c]	[VBS10a]	[SM14a]	[AKK+17]	[EDBC13]
Technology	40 nm	90 nm	250 nm	32 nm SOI	130 nm
Capacitors	MOM	MIM	MIM	Deep-trench	Ferro-electric
iVCRs	2:1	5:4 3:2	4 bit recursive	3:2 2:1	1:1 3:2 2:1 3:1
V_{in} [V]	1.855–2.07	0.7–1.2	2.5	1.8	1.5
V_{out} [V]	0.9	0.5–0.85	0.1–2.18	0.7–1.1	0.4–1.1
P_L @ η_{peak} [mW]	3.15	2.7	2.4[a]	900[a]	0.48[a]
Area [mm^2]	2.4	3	4.65	1.968	0.37
P_D @ η_{peak} [mW/mm^2]	1.3	0.9	0.52[a]	460[a]	1.3[a]
η_{peak}	94.6%	87%	85%	88%[a]	93%
η_{peak} @ iVCR = 2:1	94.6%	–	85%[a]	88%[a]	91%[a]
P_N @ iVCR = 2:1	5.7%	–	17.7%[a]	13.6%[a]	9.9%[a]
V_{ripple} [mV]	18	60	–	30	–
Number of multiphases	16	1	1	64	4
C_{DC} [nF]	0	3.7[a]	–	0	–
FoM_{LD}[b] @ η_{peak} [V^{-1}]	0.06	0.70	0.15	0.16	0.21

[a]Estimate based on graphs
[b]See (2.70), lower is better

Redistribution was proposed that reduces these parasitic coupling losses up to any desired level by continuously redistributing parasitic charge in-between phase-shifted converter cores. Because according to the usual theoretical models this could lead to efficiencies arbitrarily close to 100%, the effect of transistor leakage was investigated and was found to be another limiting factor in the maximum obtainable efficiency of a SC regulator. Theoretical analysis showed that SPCR significantly increases achievable efficiencies from infinitely low- to very high (> 10 W/mm^2) power-densities and for a wide variety of technological parameters.

Thanks to the use of charge redistribution buses, SPCR only requires little area overhead and is relatively easy to implement. A 2:1 converter was fabricated in a 40 nm bulk CMOS technology that achieved a record fully integrated SC regulator efficiency of 94.6%.

Chapter 5
MIMO Switched-Capacitor Converter Using Parasitic Coupling

During the normal operation of the SPCR technique introduced in Chap. 4, additional DC voltages are generated in the form of the charge redistribution buses. In essence, this means that the switching parasitic coupling itself has been transformed into a multiple-input multiple-output (MIMO) converter, thus using the parasitic coupling to its advantage. This Chapter will discuss how these generated DC voltages could be used within a larger switched-capacitor design, and will analyze how well the emergent parasitic MIMO converter performs relative to other MIMO SC topologies.

5.1 Motivation

While switches are often idealized in early stages of the design of an SC converter, at one point they need to be translated to transistors that require appropriate gate-driving signals. For converters with a low in- and output voltage relative to the technology's supply voltage, this can be accomplished by directly using a combination of the ground, in- and/or output voltage. With larger in- and/or output voltages, however, said combination might no longer suffice to drive the transistor's gate while also respecting its maximum voltage rating, as illustrated by Fig. 5.1. In these situations, additional voltage rails are thus a necessity, but even when they are not, they can be advantageous if they increase the power transistors' overdrive voltage. In the literature, these voltage rails are typically generated using either a separate DC-DC converter per rail [KP14], or one MIMO DC-DC converter [MPK+15]. For other applications, such multiple-output converters have also been successfully adopted [HL15, TS16, DCVBDS12], showing the promise of this technology. Alternatively, a bootstrap circuit can be used to create a voltage relative to the transistor's source node [SS15a]. Either way, the generation of extra rails

© Springer Nature Switzerland AG 2020

N. Butzen, M. Steyaert, *Advanced Multiphasing Switched-Capacitor DC-DC Converters*, https://doi.org/10.1007/978-3-030-38735-8_5

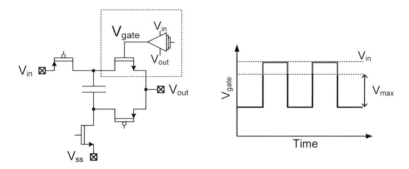

Fig. 5.1 Example gate driving scheme for a topside transistor in a 2:1 SC DC–DC converter. With large V_{in}, the transistor's voltage rating is no longer respected

takes up a substantial part of the die area, may even require additional external components [HL15, TS16], increases the total system complexity, and generally only achieves limited efficiency.

Another open research question is the overhead losses of ultra-low-power converters. Here, the output power is generally so low that even the overhead (control, clock generation, etc.) can be a bottleneck for the achievable efficiency [PC17]. This is especially problematic considering these kinds of converters are most-often used for severely energy-limited applications like energy scavenging or harvesting [JOB+14], where efficiency is the most important specification. In these situations, lowering the supply voltage of most of the overhead circuitry by using an additional voltage rail could lead to a significant reduction of both dynamic- and static power losses. Furthermore, due to the low control frequencies of low-power converters, large reductions of the supply voltage are possible before running into problems with the overhead's timing requirements. As before, however, the generation of the rail itself is a burden.

In Chap. 4, a technique was introduced that significantly increases the efficiency of SC converters by redistributing charge between parasitic capacitors. Exploring the SPCR technique, a series of DC voltage levels that are evenly spread across the flying capacitors' bottom-plate swing can easily be implemented. This Chapter proposes using these intrinsically generated DC voltages, or charge redistribution buses, to efficiently power circuits within the converter, such as the ones described above, without any area overhead or added complexity. Figure 5.2 illustrates how such an implementation could look like.

This chapter is organized as follows: Sect. 5.2 deals with the parasitic MIMO converter's basic working principle. This type of converter's characterization and regulation is discussed in Sect. 5.3, and compared to regular SC MIMO converters in Sect. 5.4. Section 5.5 discusses measurement results of the presented converter and Sect. 5.6 summarizes this work.

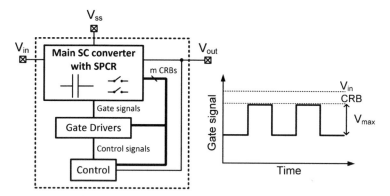

Fig. 5.2 High level schema of an example system that uses m CRBs, generated by the SPCR technique, to power gate drivers and the converter controller

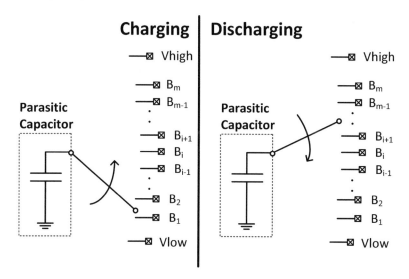

Fig. 5.3 Working principle of SPCR technique from the point of view of a single parasitic capacitor

5.2   Working Principle

In Fig. 5.3 the working principle of SPCR from the point-of-view of the parasitic capacitor is shown. Rather than continuously switching between V_{high} and V_{low}, the capacitor will instead connect to intermediate voltage rails B_1 to B_m in ascending order when charging and descending order when discharging. In order to explain how this results in a series of DC voltage rails, it helps to assume an infinite decoupling capacitor at each rail. When the parasitic capacitor, C_{par}, connects to B_i in its charging phase, an amount of charge, dependent on the difference between the rail's voltage, $V_{B,i}$, and the previous rail's voltage, $V_{B,i-1}$, is transferred. Similarly, in the discharging phase, the transferred charge depends on $(V_{B,i+1} - V_{B,i})$. Assuming B_i is unloaded, the sum of the charges of both phases must be zero when the system is in steady-state. This means that the following must be true:

$$V_{B,i} = \frac{V_{B,i+1} + V_{B,i-1}}{2}. \tag{5.1}$$

In other words, if B_i is unloaded, then its voltage will converge to the average of the adjacent nodes' voltages. This property, while simple, is key to understanding this type of MIMO converter and is used throughout this Chapter.

Now, if all intermediate nodes are unloaded, (5.1) can be used to prove that said nodes will spread out evenly between the boundary conditions set by V_{high} and V_{low}:

$$V_{B,i} = \frac{i}{m+1}\Delta V + V_{low}, \tag{5.2}$$

where ΔV is the voltage difference between V_{high} and V_{low}. Thus, from two reference voltages V_{high} and V_{low}, the parasitic converter can generate any number of desired DC nodes using nothing but the parasitic coupling of a main SC converter. Simply by adding more intermediate steps, more DC voltages are generated.

In practice, very large decoupling capacitors are impractical and costly to realize. However, it is possible to achieve the same effect without decoupling. After all, the purpose of the decoupling capacitor is simply that of a buffer: it holds a certain amount of charge until it is required again by the parasitic converter. As such, one can argue that the parasitic capacitor is exchanging charge with a time-shifted version of itself. By splitting the parasitic capacitor up into multiple phase-shifted versions of itself, each phase can at each point in time connect to another that goes through the opposite voltage step, as was done in Chap. 4 (Fig. 5.4), and the

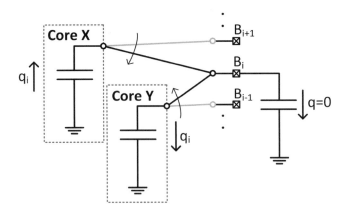

Fig. 5.4 Working principle of SPCR technique in steady-state using phase-shifted cores

middleman can be cut. Thus, when using the SPCR technique, the extra voltage rails require no decoupling and are by consequence truly generated without area overhead.

5.3 Characterization and Regulation

Before getting into further analysis, it is important to emphasize the situation for which we are analyzing this type of SC converter. In the context of a main converter that requires certain output voltages internally, the proposed converter can be used without the addition of any circuitry, which means that, unlike with a regular SC converter, non-conductive losses (parasitic coupling losses, gate charging, leakage, etc.) are not introduced. In the parasitic converter there is, on the other hand, always a transfer of charge, even when none of the generated voltage rails is loaded. The losses associated with these charge transfers, however, are the bottom-plate losses of the main converter and are thus not considered in further analysis. After all, they are already taken into account in the design of the main converter. What is investigated instead is how adding a load on one of the intermediate nodes generated by SPCR increases the total power consumption of the main converter, $P_{mainconverter}$. The change in power is considered to be the true parasitic converter input power:

$$P_{in} = \Delta P_{mainconverter}. \tag{5.3}$$

5.3.1 General MIMO SC Model

In Fig. 2.4 the typical model of a single-output SC converter is shown. In this model, infinite decoupling is assumed to be present at the output, which means that the output voltage is perfect DC, while the output current is time-averaged. Also, note that the converter in this general model consists of ports, with each port having a positive and a negative terminal. In practice, though, both ports will often share their negative terminal.

If no parasitic elements are present, only two parameters suffice to fully characterize a converter. The iVCR determines the relation between in- and output voltage, while the output resistance, R_{out}, relates the voltage drop, ΔV_{drop}, at the output, to the current that is drawn from it:

$$\Delta V_{drop} = R_{out} I. \tag{5.4}$$

The single-output converter's conductive losses are determined by the latter:

$$P_{loss} = R_{out} I^2. \tag{5.5}$$

Fig. 5.5 Model of an SC
converter with multiple ports

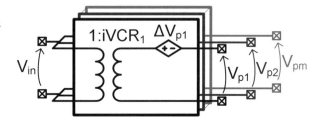

Whereas in the above discussion the terms input and output are used, from an energy perspective the input and output are not determined by the topology but by the sign of the time-averaged current. For example, if the time-averaged current in (5.4) changes sign, what was considered the output will in fact be sourcing charge and thus current.

When generalizing this model to multiple ports it becomes apparent that the complexity cannot scale linearly with the number of ports, m. This is due to the fact that a load current supplied to a port can influence the voltage across every other port's terminals. However, it still does so in linear fashion. Figure 5.5 shows the model proposed in [DLAH14]. Here, the voltage drop across the terminals of port i, ΔV_{pi}, is shown to be

$$\Delta V_{pi} = \sum_{j=1}^{m} z_{ij} I_{pj}, \tag{5.6}$$

where I_{pj} is the time-averaged load current supplied to port j, and z_{ij} is the transimpedance from the time-averaged current supplied to port j to the voltage across the terminals of port i. The full system can consequently be described by a simple algebraic equation

$$\overrightarrow{V_p} = \overrightarrow{iVCR}\, V_{in} - \mathbf{Z}\overrightarrow{I_p}, \tag{5.7}$$

with V_{in} the voltage across the input port's terminals as shown in Fig. 5.5, $\overrightarrow{V_p}$, $\overrightarrow{iVCR}$ and $\overrightarrow{I_p}$ vectors containing the respective voltage, conversion ratio, and time-averaged current for each port, and $\mathbf{Z}$ the converter's impedance matrix, whose elements are the transimpedances used in (5.6). Similar to the single-output case, the total conduction losses can be written as

$$P_{loss} = \overrightarrow{I_p}^T \mathbf{Z} \overrightarrow{I_p}. \tag{5.8}$$

Note that, similar to the two-port case, each port can source or sink current to the system, depending on the sign of its corresponding current. As such, this model describes MIMO operation.

5.3.2 *Characterization*

Within the context of a main converter using SPCR, the switches that connect the parasitic capacitor to the different intermediate nodes should be large enough to allow close to full settling of the capacitor's voltage to enable the full reduction in bottom-plate losses. Therefore, the parasitic converter is assumed to operate in the slow-switching limit (SSL) [See09], where the switch resistance can be neglected.

In [DLAH14] a method is described to work out $\mathbf{Z}$ given an SC converter's topology by using a charge-based analysis for each node separately. Due to the highly regular structure of the parasitic MIMO converter, however, it is possible to derive the elements of $\mathbf{Z}$ directly using a more intuitive approach.

Consider an example parasitic converter with 6 intermediate nodes. Each node is the positive terminal of a port, with the corresponding negative terminal being V_{low} for all ports. Figure 5.6 illustrates how the node voltages shift when a single load is introduced. For all nodes without a load, the relation described by Eq. (5.1) still holds. At the loaded node, on the other hand, a certain amount of charge flows to the load each clock cycle, and will pull said node down. Thus, the load changes the voltage differences between the nodes. Above the loaded node, the voltage difference, ΔV_{above}, will be enlarged, while the opposite is true for the voltage difference of the bottom nodes, ΔV_{below}. Using the condition of charge preservation in steady-state at the loaded node, j, the relation between both can be written as

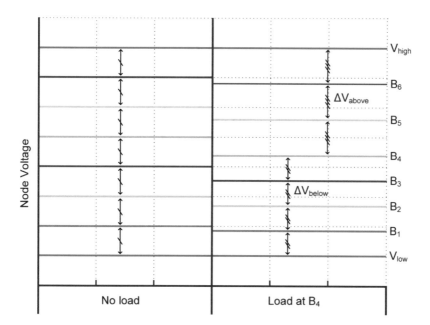

Fig. 5.6 Waveforms of the node voltages of a 6 node parasitic MIMO SC converter under no-load-and loading conditions

$$\Delta V_{above} = \Delta V_{below} + \frac{q_{out,j}}{C_{tot}}, \qquad (5.9)$$

where $q_{out,j}$ is the load charge over an entire clock period and C_{tot} is the total capacitance of the parasitic converter. Regardless of the size of the load, the sum of all voltage differences must, of course, still be equal to the total ΔV. By using this fact together with (5.9), the following can be derived:

$$\Delta V_{above} = \frac{\Delta V}{m+1} + \frac{j}{m+1}\frac{q_{out,j}}{C_{tot}}, \qquad (5.10)$$

$$\Delta V_{below} = \frac{\Delta V}{m+1} + \left(\frac{j}{m+1} - 1\right)\frac{q_{out,j}}{C_{tot}}, \qquad (5.11)$$

which in turn can be used to determine the voltage of any node, i,

$$V_{B,i} = \begin{cases} V_{low} + \frac{i}{m+1}\Delta V + j\left(\frac{i}{m+1} - 1\right)\frac{q_{out,j}}{C_{tot}} & i > j \\[3mm] V_{low} + \frac{i}{m+1}\Delta V + i\left(\frac{j}{m+1} - 1\right)\frac{q_{out,j}}{C_{tot}} & i \le j. \end{cases} \qquad (5.12)$$

Considering the fact that the time-averaged load current drawn from a node equals $f_{sw}q_{out,j}$, where f_{sw} is the converter frequency corresponding to a full clock cycle, all elements of the impedance matrix $\mathbf{Z}$ can subsequently be derived:

$$z_{ij} = \begin{cases} j\left(1 - \frac{i}{m+1}\right)\frac{1}{f_{sw}C_{tot}} & i > j \\[3mm] i\left(1 - \frac{j}{m+1}\right)\frac{1}{f_{sw}C_{tot}} & i \le j. \end{cases} \qquad (5.13)$$

Table 5.1 shows the impedance matrix for a select number of output nodes.

Table 5.1 Parasitic converter's normalized impedance matrix for varying number of output nodes

m	1	2	3	4
$f_{sw}C_{tot}\mathbf{Z}$	$\dfrac{1}{2}$	$\dfrac{1}{3}\begin{bmatrix} 2 & 1 \\ 1 & 2 \end{bmatrix}$	$\dfrac{1}{4}\begin{bmatrix} 3 & 2 & 1 \\ 2 & 4 & 2 \\ 1 & 2 & 3 \end{bmatrix}$	$\dfrac{1}{5}\begin{bmatrix} 4 & 3 & 2 & 1 \\ 3 & 6 & 4 & 2 \\ 2 & 4 & 6 & 3 \\ 1 & 2 & 3 & 4 \end{bmatrix}$

5.3.3 Natural Regulation

The parasitic converter's frequency is set by the frequency of the main converter. At first glance, it might seem like this makes regulation of the parasitic converter's intermediate nodes difficult to achieve. In practice, though, this inherent sharing of working frequency leads to an interesting benefit of the proposed converter which is referred to as *natural regulation*.

In a SC converter, there are many overhead loss contributors whose power consumption scales largely linearly with the converter's switching frequency. Such contributors include gate driver, phase generators, interconnect buffers, and so on. When these blocks are powered by the intermediate voltage rails, they will only require energy when the main converter is switching, as pointed out in [MPK+15]. In the presented case, however, energy will be transferred to the intermediate rails whenever the main converter switches by means of the parasitic converter. As such, regulation is achieved naturally without overhead.

For overhead loss contributors that demand energy regardless of the main converter's switching frequency, e.g. a hysteretic controller's comparator, the main converter will need to realize a minimal switching frequency in order to meet the minimum voltage requirements for these blocks. This situation, though, is not very different from that of a converter whose control is powered by its own output voltage. Also here, the converter will be switching every so often to power its overhead circuitry.

5.4 Topology Comparison

In this section, the presented parasitic MIMO topology is compared to two regular SC MIMO topologies based on the commonly-used Ladder- [MPK+15, LTZ+15] and Dickson topologies, shown in Fig. 5.7a, b, respectively. The negative terminal of all ports is assumed to be V_{low}. By consequence, the discussion can be simplified to one of the positive terminals or nodes.

5.4.1 Losses

A switched-capacitor converter's losses can be divided into two groups: Conductive losses due to its finite output resistance and non-conductive losses such as power transistor leakage, gate-charging losses, and parasitic substrate coupling losses. Because the parasitic converter does not add any power transistors, nor flying capacitance to the system, it will neither add leakage-, gate-charging-, and parasitic substrate coupling losses. As such it has a significant head-start compared to regular SC topologies.

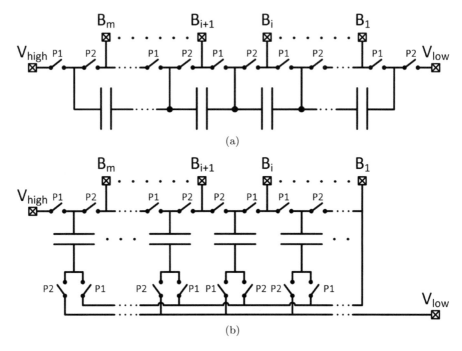

Fig. 5.7 Generalized topology of the (**a**) Ladder- and (**b**) Dickson MIMO converters

The output resistance of a SC converter has an SSL and FSL component [See09]. As established in Sect. 5.3.2, due to the sizing of the switches of the parasitic converter, it is assumed to work in the SSL regime, where the FSL-component can be neglected. As such, the comparison here focuses on the SSL-resistance.

For a single-output converter, different topologies can be evaluated by simply comparing their SSL topology factors, K_C or K'_C, to one another [VBS10b]. These are in essence their SSL-resistance, but normalized for switch frequency and total flying capacitance or energy, as discussed in Chap. 2. After all, a topology with a larger output resistance will have larger conductive losses for the same load current. As such, single-output converters can be ordered in terms of SSL performance, regardless of the nature of the load.

For MIMO converters, on the other hand, evaluation is not as clear-cut. As demonstrated by (5.8), the conductive losses depend on $\vec{I}_p$, which means that, even after normalization, the relative sizes of the currents drawn from the nodes have a significant impact on a direct comparison. As a result, while the size of the elements of the impedance matrix, $\mathbf{Z}$, can give an initial impression, any general comparison of topologies must also include multiple specific use cases for the converter.

Element Based

Figure 5.8 gives a first element-based comparison of the different topologies for 7 output nodes. Similar to the single-output case, the impedance elements are normalized by multiplying them with the switching frequency and total capacitance. The Dickson topology has the lowest normalized impedance of all converters in output 1, which is the node with the lowest voltage. This result matches earlier single-output analysis [See09]. For every other node, however, the parasitic converter has significantly lower output impedance: four to six times better compared to the Dickson-, and two orders of magnitude better compared to the Ladder converter.

Use Case Based

The topologies' normalized conductive losses are compared for three different node current profiles, each of which corresponds to a specific use case. Table 5.2 gives an overview of the used current profiles, which are, unlike in [BS16b], further

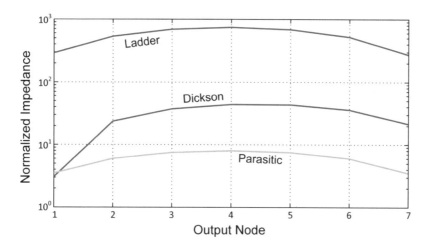

Fig. 5.8 Comparison of the sum of normalized output impedances per output node for different SC MIMO converters with 7 nodes

Table 5.2 Overview of current profiles	Use case	$\overrightarrow{I_p}$
	Iso-current	$\frac{2}{m}\begin{bmatrix} 1 & 1 & \cdots & 1 \end{bmatrix}^T$
	Iso-power	$\frac{m+1}{m}\begin{bmatrix} 1 & \frac{1}{2} & \cdots & \frac{1}{m} \end{bmatrix}^T$
	Single-output iVCR $= 0.5$	$\begin{bmatrix} 0 & \cdots & 0 & 2 & 0 & \cdots & 0 \end{bmatrix}^T$

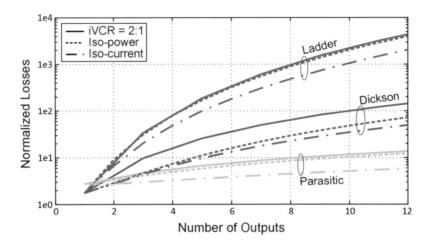

Fig. 5.9 Comparison of the normalized conduction losses of the parasitic-, to the ladder- and Dickson MIMO SC converter for an equal load current drawn from each node, a load current drawn from each node inversely proportional to the iVCR and a single load current drawn from an iVCR of 2:1

normalized such that the total output power in each case is approximately the same. These consist of two multiple-output cases and one where the MIMO converter is used to supply a single voltage corresponding to an iVCR of 2:1. The results of this comparison are shown in Fig. 5.9. When used with 2 or fewer outputs, the parasitic converter shows higher losses than the Dickson converter. For a larger number of outputs, on the other hand, the regular converters see a rapid increase in losses. The relative gain of the proposed converter gets consequently larger for increasing number of outputs. Moreover, this statement holds regardless of the use-case for which the topologies are compared. Thus it can be concluded that the parasitic converter has exceptionally low losses.

5.4.2 Transient Behavior

The transient behavior of the proposed-, Ladder-, and Dickson topologies is compared for an implementation with nine intermediate nodes. In order to be able to remove the load capacitance of every intermediate node, the ladder- and Dickson converter are implemented as two converter cores which run in anti-phase. This way there is always at least one capacitor connected to every node. In addition, the switch resistance is assumed to be close to zero such that all converters run fully in the SSL region.

Figure 5.10 shows the simulated transient response of the intermediate nodes corresponding to an iVCR of 10:9 and 2:1, when a load step is applied to the intermediate node with an iVCR of 10:9. Both in cross- (Fig. 5.10a) and self-

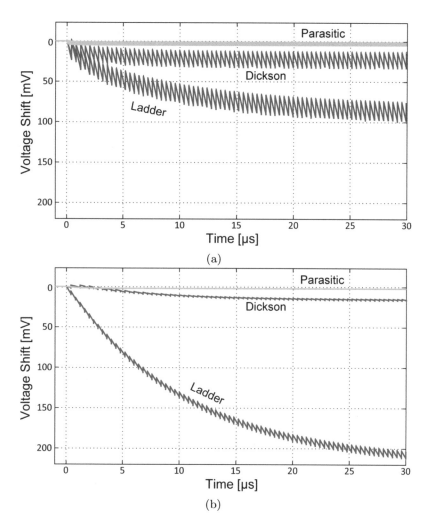

Fig. 5.10 Comparison of the simulated voltage shift waveforms of the intermediate nodes corresponding to an iVCR of (**a**) 10:9 and (**b**) 2:1, for a parasitic-, Ladder-, and Dickson-converter with a total of 9 intermediate nodes, when a load step of 500 nA is applied to the intermediate node with an iVCR of 10:9 at $t = 0$ s. Here, the total flying capacitance is 100 pF and $f_{sw} = 1.25$ MHz

regulation (Fig. 5.10b) situations the parasitic converter undergoes the smallest voltage deviation at 2 and 2.9 mV, respectively, due to its low output impedance. The Dickson- with a respective deviation of 15 and 23 mV, and the Ladder-converter, with a 207 and 87 mV drop, respectively, do notably worse.

In addition, the parasitic converter has significantly lower ripple. Under self-regulation, for example, its ripple is only 2.2 mV, compared to 18.7 and 23.9 mV for the Dickson- and Ladder, converter respectively. When a load is applied to a node, the other intermediate nodes of the parasitic converter show no ripple. This is due to the fact that at no point in time a capacitor of the parasitic converter connects

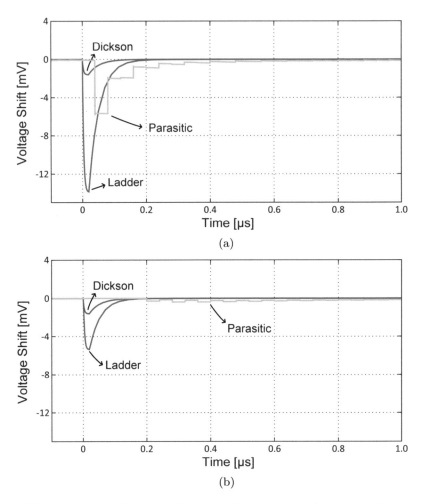

Fig. 5.11 Comparison of the simulated voltage shift waveforms of the intermediate nodes corresponding to an iVCR of (**a**) 10:9 and (**b**) 2:1, for a parasitic-, Ladder- and Dickson-converter with a total of 9 intermediate nodes, when a droop is applied to the input at $t = 0$ s, representative of the droop reported in Chap. 4. Here, the total flying capacitance is 100 pF and $f_{sw} = 1.25$ MHz

to more than one node at the same time. As such, no low-impedance current path exists between any pair of nodes, leading to a high degree of isolation.

Figure 5.11 compares the line regulation of the same intermediate nodes, under a line droop of 21 mV that settles within 120 ns, which is representative of the droop of the main converter in Chap. 4. At the intermediate node corresponding to an iVCR of 10:9, the Dickson converter has the smallest droop of 1.6 mV. The ladder- and parasitic converter, on the other hand, have droops of 13.9 and 5.7 mV, respectively. It is worth noticing that the parasitic converter has a delayed response to the line droop. This is due to the earlier discussed isolation between the intermediate nodes, which also extends to the V_{high} node: It is only when the core which was connected to V_{high} switches to the aforementioned intermediate node, that said node's voltage deviates. Furthermore, in doing so it connects to a core that has not yet connected to a perturbed node, leading to the voltage deviation being averaged out. This also means by extension that intermediate nodes with smaller iVCRs (10:8, 10:7, ...) have increasingly better isolation from line transients because a core goes through more averaging steps before connecting with said nodes. This is demonstrated by Fig. 5.10b. As can be seen, the node corresponding to an iVCR of 2:1 has a significantly delayed response and a voltage perturbation less than 0.4 mV.

5.5 Verification

5.5.1 Measurement Setup

To verify the working principle and qualitative analysis of the parasitic converter, measurements are performed on the main converter discussed in Chap. 4. A schematic representation of the measurement setup of the parasitic converter within the main converter is given in Fig. 5.12. A significant part has remained unchanged

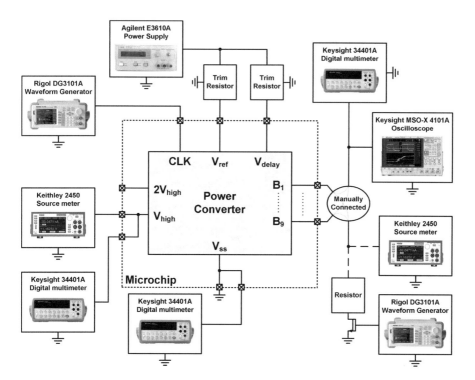

Fig. 5.12 Overview of the MIMO parasitic SC converter measurement setup

from Sect. 4.5.1. The converter uses a total of 16 out-of-phase cores to generate 9 output nodes between 900 mV and ground. The B_5 node, corresponding to an iVCR of 2:1, however, was not implemented as a physical voltage rail in this realization. None of the nodes has any decoupling added to it.

All measurements are performed by connecting a 900 mV voltage, generated with a Keithley 2450 source meter [Kei11], to V_{high}, which corresponds to the main converter's output. Its input, on the other hand, is left unconnected and its lower-bound hysteretic controller is bypassed by applying a reference voltage, V_{ref}, of approximately 1 V. This way, the main converter is unloaded and free-running, causing its influence on measurements to be reduced as much as possible. The measurements at the intermediate nodes are done by loading a node with a source meter while measuring its (or another node's) voltage with a Keysight 34401A multimeter [Key12], or a Keysight MSO-X 4101A oscilloscope [Key17]. For transient measurements, a switched resistor, activated using a series-connected, low on-resistance MOSFET that is in turn driven with a Rigol DG3101A waveform generator [RIG08], is used instead.

5.5.2 Output Impedance

Figure 5.13 shows the measured voltage of the DC voltage rails versus the rail current, in both source and sink conditions. The dropout voltage of the parasitic converter's nodes scales linearly with load current, as demonstrated in Sect. 5.3.1. The largest and smallest absolute voltage drop at the largest measured load current

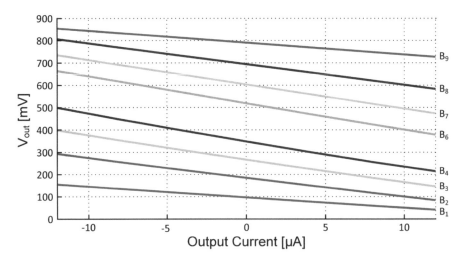

Fig. 5.13 Measured output voltage versus output current for all DC voltage rails with $f_{sw} = 1.6$ MHz and $V_{high} = 900$ mV

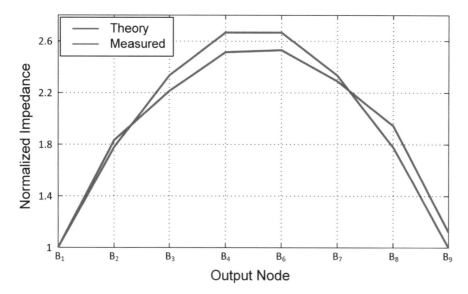

Fig. 5.14 Comparison of relative size of the measured output impedance of all DC nodes to theoretical predictions

of $12\,\mu A$ are 141 and 55 mV for B_6 and B_9, respectively, which is consistent with the findings of Sect. 5.3.2. The parasitic converter's output impedance is calculated using the measured voltage deviation under loading and compared to the theoretical expected value based on the analysis of Sect. 5.3.2 in Fig. 5.14. A good match can be witnessed between both, with a maximum relative error of 10% on B_9.

To demonstrate the MIMO nature of the presented parasitic converter, the influence on the voltage of one node due to a load current on a different node is measured. These results are shown in Fig. 5.15 using B_9. As expected, the nodes closest to B_9 have the most impact on its voltage, causing a shift of up to 50 mV at $12\,\mu A$.

5.5.3   Transient Measurement

In Fig. 5.16 the measured cross-regulation of the parasitic converter is shown. Here, a current step is applied to B_9, leading to voltage shift of 50 mV with a transient time of 50 ns. The intermediate nodes B_8 and B_6 shift 18 and 8 mV, respectively, as a result, and settle within approximately $1\,\mu s$. No notable effect on B_2 is measured. As expected, the intermediate nodes closest to the perturbed node have the largest response.

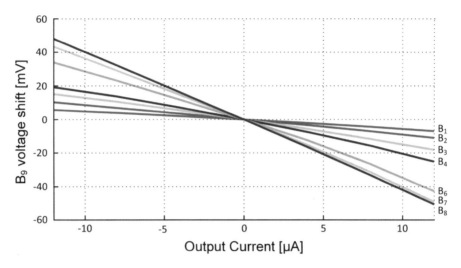

Fig. 5.15 Measured voltage shift on B_9 under the influence of a load current drawn from B_1 to B_8

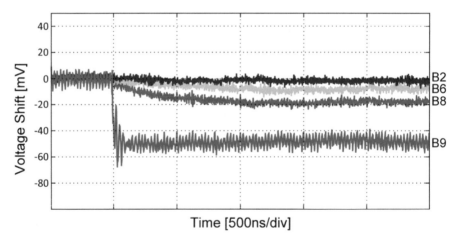

Fig. 5.16 Measured voltage waveforms of intermediate nodes B_2, B_6, B_8, and B_9, when a voltage step of 50 mV is applied to B_9 with a transient time of 50 ns

5.5.4 Efficiency

In order to measure the efficiency of the parasitic MIMO converter, first a reference measurement is performed to determine the converter's current at V_{high} under no-loading operation. By then measuring the supply current when a load current is present on one of the DC rails, and comparing said current to the reference measurement, the input power of the parasitic converter can be calculated using (5.3).

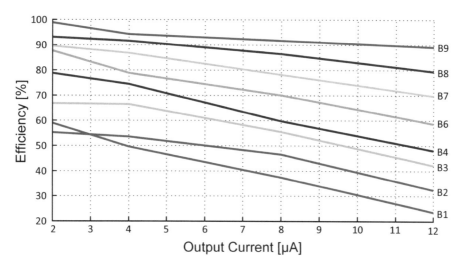

Fig. 5.17 Measured efficiency versus output current for B_1 to B_9 with $f_{sw} = 1.6\,\text{MHz}$ and $V_{high} = 900\,\text{mV}$

$$\eta_{node} = \frac{P_{node}}{\Delta P_{mainconverter}}. \tag{5.14}$$

Figure 5.17 shows the resulting measured efficiency of the presented converter. Due to the conductive nature of the losses, the efficiency is largely dependent on the relative size of the load impedance to the equivalent output impedance of the converter, which means that at constant output current, a lower voltage leads to lower efficiencies. Moreover, as load current tend towards zero, the efficiency approaches 100%. The measured peak efficiency is 98.9% at a load current of $2\,\mu\text{A}$ at B_9. Lower load currents would have yielded even higher efficiencies, but due to the fact that the input power is calculated by subtracting two larger numbers, these measurements could not be performed with the required accuracy. The highest output power of $8.7\,\mu\text{W}$ is also measured at B_9 with an efficiency of 89%. According to simulations, this power is similar to the power consumption of the main converter's controller, which is further testament to the possible use of these voltage rails within the larger converter itself.

5.5.5 Overview

Finally, an overview of the realized parasitic converter is given in Table 5.3. Due to the fact that the proposed converter makes use of the parasitic substrate coupling already present in a main SC converter, it introduces no additional die area. Moreover, this means that the converter only has conductive losses, leading

Table 5.3 Overview of converter specifications

Item	This work [BS17c]
Technology	40 nm
Capacitors	Parasitic coupling
Application	Internal SC blocks
V_{in} [V]	0.9
Outputs	8
V_{out}'s [V]	0.09, 0.18, 0.27, 0.36, 0.54, 0.63, 0.72, 0.81
η_{peak}	98.9%[a]
System η	94.6%
Max P_{out}	8.7 μW
Added die area	0 mm^2 [a]
Closed-loop?	No
Requirements	Main SC converter with SPCR

[a]Effective

to an effective efficiency of 98.9%. The downside, naturally, is that its application scope is limited to the auxiliary converter use case discussed in this work.

5.6   Conclusion

In this Chapter the challenges with switched-capacitor converter's overhead losses and gate driving were discussed together with the problems associated with internal voltage rail generation. A type of multiple-input multiple-output SC converter was introduced that generates multiple DC voltages using only the parasitic capacitance already present in fully integrated SC converters. When used in a SC converter together with Scalable Parasitic Charge Redistribution, these DC voltages can be used to power low-power control blocks or gate drivers. Furthermore, because this is achieved without adding any area overhead, no additional parasitic losses (bottom-plate losses, gate-charging losses,...) are introduced. Moreover, the conductive nature of the remaining losses means that efficiencies close to 100% can be achieved for low output powers.

A model for this type of MIMO converter was proposed and characterized, which was then used to compare the converter to known regular SC MIMO converters for different use cases. Especially for larger number of input-or output nodes, the presented converter was shown to have particularly low losses and excellent cross- and line-regulation. The basic working principle of the presented converter was verified using measurements, showing a good match between the measured and theoretical output impedance. In addition, a peak efficiency of 98.9% and output powers sufficient to power internal converter blocks were witnessed.

Chapter 6
Stage Outphasing and Multiphase Soft-Charging

While Chap. 4 introduced a method to reduce parasitic coupling losses that provided gains up to higher power-densities, the effectiveness hereof does drop substantially at these high power-densities. Naturally, this is to be expected due to the lesser importance of parasitic coupling losses in this regime, as discussed in Chap. 2.

In this chapter, another two advanced multiphasing techniques are proposed, called Stage Outphasing and Multiphase Soft-Charging, that both aim to spread the charge transfers between flying capacitors out over time. The capacitive conduction losses, which remain important over the entire power-density design space, are consequently reduced. An implementation of a 3:1 Dickson converter using these techniques verifies their working principle and achieves a record 82%-efficiency 1.1 W/mm²-power-density combination in a baseline 28 nm CMOS technology.

6.1 Background and Motivation

Monolithic switched-capacitor converters have the potential to significantly reduce PDN related losses in smartphone- and wearables' SoCs. In these applications, the converter does not only need to be efficient, but also needs to have a large iVCR, and a sufficiently high power-density. An often quoted figure is that a monolithic power converter should be able to provide the required output power in less than 10% of the die area [SAT16], which is approximately the portion of area used for decoupling today [SM15]. For smartphone SoCs, this corresponds to a power-density of roughly 500 mW/mm².

Chapter 2 revealed how, at low power-densities, the converter's efficiency is limited by the ratio of the flying capacitor's parasitic substrate coupling and the flying capacitors' capacitance or energy density. For increasingly large output power-densities, on the other hand, the switched-capacitor converter's design space has a clear efficiency-power-density trade-off. Here, it is the ratio of the transistors'

© Springer Nature Switzerland AG 2020

N. Butzen, M. Steyaert, *Advanced Multiphasing Switched-Capacitor DC-DC Converters*, https://doi.org/10.1007/978-3-030-38735-8_6

$E_{on}R_{on}$ or E_{on}/A_{on}, and the capacitor energy density that are the dominant technology factors. Either way, the capacitance density plays a crucial role in both regimes.

Due to the inherent planar nature of modern technology nodes, the energy density of capacitors is quite low and has been proven to be a severe burden, causing designs in literature to either achieve high efficiency at low power-density, like in Chap. 4 (94.6% at 1.3 mW/mm^2), or low efficiency at high power-densities [MPK$^+$15, SSP$^+$09] (60% at 1.1 W/mm^2, 0.77 W/mm^2, respectively). Converters using SOI or Deep-Trench capacitors have improved transistors or >20x higher capacitance density at their disposal, causing them to perform significantly better as a result [LSA11, AKK$^+$14], but are more expensive and/or rely on devices that are far from common. Moreover, all of these converters obtain their best result at an iVCR of only 3:2 or 2:1, leaving much room to reduce PDN losses further.

In [CKP16] a topology is proposed which, while it does not extend on multi-phasing, increases the effective flying capacitance density through soft-charging, subsequently leading to a boost for both high-and low power-densities. Said topology is, however, based on a Series-Parallel topology which is inherently less efficient at low power-densities, and the relative improvement is limited to a factor of two.

This chapter is organized as follows. In Sect. 6.2, two soft-charging techniques that increase the effective flying capacitance, called Stage Outphasing (SO) and Multiphase Soft-Charging (MSC), are introduced. Section 6.3 goes into more detail how these techniques can be efficiently implemented using an example implementation. Measurement results, verifying the effectiveness of SO and MSC in a realized converter, are discussed in Sect. 6.4. Finally, Sect. 6.5 highlights the important conclusions of this chapter in a brief summary.

6.2 Working Principle

While the proposed techniques and concepts can be applied to multiple SC topologies, they will first be explained by their use in a Dickson converter, shown in Fig. 6.1. This is simply due to the fact that the Dickson converter is a very common

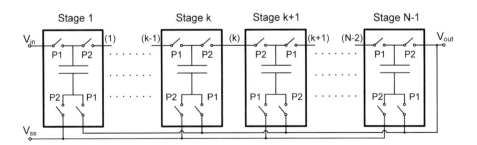

Fig. 6.1 A regular N:1 Dickson SC converter

topology which does not need any modifications to implement the techniques. An additional advantage of the Dickson topology is that it can be extended to gearbox or multi-ratio use with relative ease thanks to the Folding concept [SS15a]. Both techniques presented in this chapter are fully compatible with said concept.

6.2.1 Regular Converter

In a regular N:1 Dickson converter, there are a total of N-1 stages. If V_{out} is close to the technology's supply voltage, the top-side switches are usually implemented as two stacked transistors to avoid using less-efficient I/O devices. This also leads to the creation of intermediate nodes (k), which are topologically speaking DC nodes and can be used as voltage rails.

All stages of a Dickson converter are identical, except for the fact that they are shifted in the voltage domain, and that adjacent stages run in anti-phase of one another. Thanks to this symmetry, the analysis of a Dickson converter can be simplified by looking at just a single stage's charging- and discharging state, and extrapolating these results to the rest of the converter. Alternatively, because of the same symmetry discussed above, it is possible to look at the charge transfer(s) between two arbitrary stages. After all, in a single charge transfer, one stage is charged and another is discharged, giving the same information as the analysis of just one stage.

Figure 6.2 portrays such a charge transfer. Here a certain amount of charge, q is transferred from stage k to stage $k+1$. This charge transfer generates a voltage swing across both capacitor's terminals, ΔV, which is proportionate to q, and causes charge-sharing losses P_{cs}:

$$P_{cs} \propto \frac{q^2}{C_k}. \qquad (6.1)$$

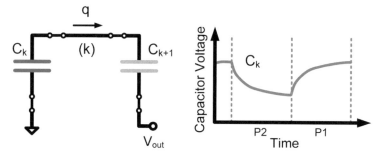

Fig. 6.2 Charge transfer between two adjacent stages of a regular Dickson converter. C_k and C_{k+1} are the flying capacitances of stages k and $k+1$, respectively, q is the transferred charge

6.2.2 Stage Outphasing

With SO, each stage is divided into two cells that connect to the same nodes, $(k-1)$ and (k), but run in anti-phase, as shown in Fig. 6.3. This way, for every stage, there is always one (and exactly one) cell that connects to each of the adjacent nodes. By phase-shifting adjacent stages 90°, a single cell connects to two cells of the adjacent stage over the duration of a charging- or discharging state, splitting said state up into two phases, as depicted in Fig. 6.4. The key here is that, in steady-state, the charges that are transferred in each phase, q_1 and q_2, are equal in size. This can be explained as follows. Let V_1 and V_3 be the top-plate voltages of C_k^1 and C_{k+1}^1, respectively, at the start of phase one, and V_2 be their TP voltage at the end of the same phase. Under the assumption that C_k^1 and C_{k+1}^1 are equal in size, V_C satisfies

$$V_1 - V_2 = V_2 - V_3. \tag{6.2}$$

Now, if the system is in steady-state and the cells of each stage are phase-shifted versions of each other, the TP voltage of C_{k+1}^2 at the end of phase 2 will also be V_3. Furthermore, because the TP of C_{k+1}^2 is connected to the TP of C_k^1 in this phase, V_3 also corresponds to the convergence voltage of the latter. From (6.2) then follows that $q_1 = q_2$. For the same total charge transferred $q = 2q_1$, the charge-sharing losses are described by

$$P_{cs} \propto 2C_k \left(\frac{\Delta V}{2}\right)^2$$

$$P_{cs} \propto \frac{q^2}{2C_k}, \tag{6.3}$$

which is a factor of 2 improvement compared to the regular Dickson case. Alternatively, twice the charge can be transferred at the same conduction efficiency, equivalent to doubling the flying capacitance.

Note that while SO does decrease the charge-sharing losses, the intermediate nodes, (k), are not topologically DC. This means that any parasitic capacitance (e.g.

Fig. 6.3 Application of Stage Outphasing to a Dickson converter

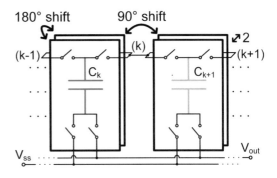

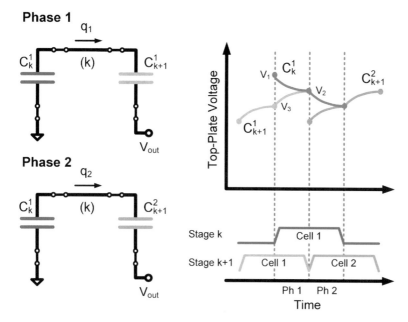

Fig. 6.4 Charge transfers between two adjacent stages of a Dickson converter using Stage Outphasing

well capacitance) on these nodes will introduce losses due to their inherent voltage swing. In practice, however, the voltage swing and capacitance on these nodes are small enough that these losses can be neglected.

Also, SO does not add any additional power transistors to the topology. In this sense, at least from a topology perspective, the gain that SO provides comes with no notable downsides. This is due to the fact that there is already some redundancy present in the form of the stacked top-side transistors in the regular topology.

What might not be obvious from the above discussion is *why* stage outphasing works. To answer this, consider the state diagram of a stage outphasing converter shown in Fig. 6.5. As can be seen, a cell in the first phase of the charging state, which has presumably the lowest voltage of all charging cells, connects to the cell in the last phase of the discharging state which in turn has the lowest voltage of all discharging cells. Similarly, the discharging cell of the first phase connects to the charging cell of the last phase, both cells having the highest voltage of their respective state. In other words, each charge transfer takes place between cells whose voltages match each other's most closely, thus minimizing the charge-sharing losses.

Note that this explanation is very similar to the explanation of the SPCR technique in Chap. 4. And indeed, when comparing Fig. 4.4 to Fig. 6.5 it can be appreciated that the underlying principle is in fact the same. The difference is that

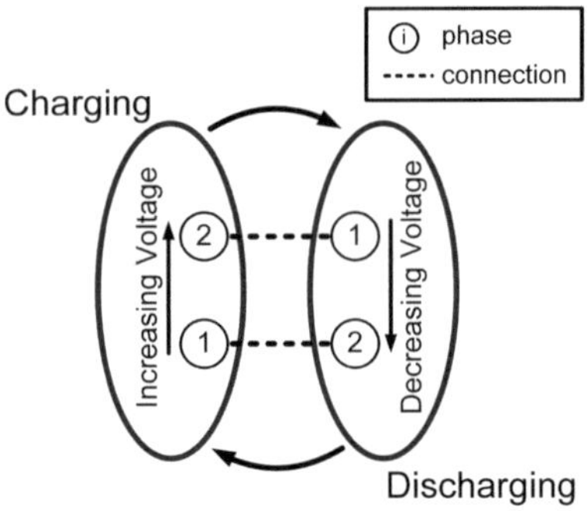

Fig. 6.5 State diagram of a
Stage Outphasing Dickson
converter

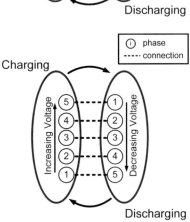

Fig. 6.6 State diagram of a
Dickson converter with 5
soft-charging steps

with SPCR, the impedance of the switching (parasitic) capacitor was increased, while here it used to *decrease* a switched-capacitor's effective impedance.

6.2.3 Multiphase Soft-Charging

Now that the working principle of stage outphasing has been related to SPCR, it is straightforward to conceive a state diagram which has even lower charge-sharing losses. Simply by adding more phases to both the charging- and discharging state, the total charge transfer will take place in more steps. An example hereof with 5 steps is shown in Fig. 6.6.

To implement this, the number of phases of the converter itself needs to be increased. This is precisely what multiphase soft-charging does, shown in Fig. 6.7. Here, each intermediate node is split into M seperate nodes, $(k, 1)$ to (k, M), and a switch is added from each cell to all adjacent nodes. In addition, each stage is split into M cells which are phase-shifted $360°/M$ and adjacent stages are further shifted $180°/M$. A full cycle takes a total of $2M$ phases which are split evenly in the charging- and discharging state.

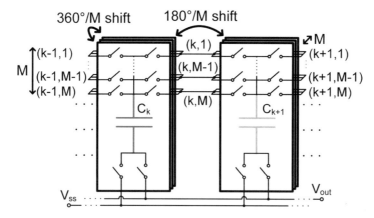

Fig. 6.7 Application of Multiphase Soft-Charging to a Dickson converter with a factor of M

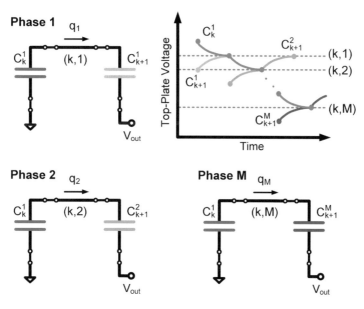

Fig. 6.8 Charge transfers between two adjacent stages of a Dickson converter using Multiphase Soft-Charging

Figure 6.8 depicts the charge transfers of a discharging cell in a converter using MSC. Using the same reasoning as in Sect. 6.2.2, it can be shown that in steady-state the charge transferred in two adjacent phases is equal in size. Extrapolating this result, the charge transfers of all phases in a state must also be equal. Moreover, this implies that the voltages of nodes $(k, 1)$ to (k, M) are equidistant. Normalizing for the total charge transferred, the total charge-sharing losses of a flying capacitor over a full charge/discharge are consequently proportionate to

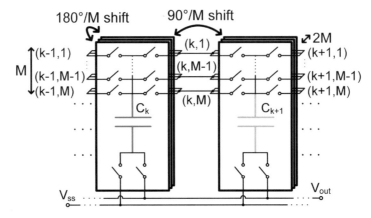

Fig. 6.9 Application of a combination of Stage Outphasing and Multiphase Soft-Charging to a Dickson converter with a factor of M

$$P_{cs} \propto \frac{q^2}{MC_k}. \tag{6.4}$$

Thus, MSC enables a scalable improvement of the charge-sharing losses by changing the factor M.

Stage outphasing and multiphase soft-charging are not mutually exclusive and can be combined with little effort. As shown in Fig. 6.9, compared to a design using only MSC, the number of cells per stage simply needs to be doubled, and the phase shift between cells of the same stage and stages themselves halved. As one would expect, the following then holds

$$P_{cs} \propto \frac{q^2}{2MC_k}. \tag{6.5}$$

6.2.4   Full Soft-Charging Converter

By definition, a transfer between a single capacitor and a fixed DC voltage (e.g. $V_{out} - V_{ss}$ or $V_{in} - V_{out}$) cannot be soft-charged. Both SO and MSC can thus only be applied to charge transfers between capacitors. For a Dickson converter, though, all but two transfers can be soft-charged, regardless of the number of stages. Because the number of stages determines the iVCR, applying SO and MSC to the Dickson topology consequently gets better for increasing iVCR.

Both SO and MSC are techniques that in essence modify a two-phase SC topology to a more efficient version using multiple phases. In Chap. 2, it was established that the performance of a SC topology is determined by its topological factors. The techniques presented here have an influence on the K_C/K_C' and K_S/K_S'

parameters. The smaller these parameters, the more efficient the topology utilizes its capacitors or switches. Because the goal is to demonstrate the relative improvement of SO and MSC, this part will simplify the analysis by only focusing on K_C and K'_S.

To derive K_C for a Dickson converter using SO and/or MSC, there are multiple options. A first one is to use the general multiphase SC model described in [DLAH14], which has the downside of requiring a lengthy calculation. A much more simple and elegant way is to start with the result for a regular $N : 1$ Dickson converter with equally, and thus optimally [See09], sized capacitors:

$$K_{C,Dickson,regular} = (1 - N^{-1})^2. \tag{6.6}$$

Assume a Dickson converter with S soft-charging phases for all capacitor-capacitor charge transfers, which means either MSC is used with $M = S$ or a combination of SO and MSC is used with $2M = S$. In this case, two charge transfers, the charging of the first stage by $V_{in} - V_{out}$ and the discharging of the final stage to $V_{out} - V_{ss}$ remain unchanged, while the $2(N-2)$ other transfers' losses are reduced by a factor of S. Keeping in mind that there are a total of $2(N-1)$ charge transfers for a regular $N : 1$ Dickson converter, each of which contributes equally to $K_{C,Dickson,regular}$, the soft-charged topological factor can be written as

$$K_{C,Dickson,sc} = \frac{K_{C,Dickson,regular}}{2(N-1)}\left(2 + \frac{2(N-2)}{S}\right) \tag{6.7}$$

$$K_{C,Dickson,sc} = \frac{N-1}{N^2}\left(1 + \frac{(N-2)}{S}\right). \tag{6.8}$$

Figure 6.10 shows how K_C scales with respect to N and the soft-charging factor, S, and includes a comparison to the regular Dickson converter. For increasingly large N, K_c tends towards S^{-1}. Furthermore, the largest relative gains are to be had for smaller S. As expected though, these gains do get more significant for larger iVCRs.

While SO does not add any power transistors, MSC does require them, which will lead to an increase of the topology's K'_S. Using the model of [DLAH14], a general solution for K'_S was found using optimal relative sizing of the switches:

$$K'_{S,Dickson,msc} = \frac{4}{N}\left(2\sqrt{M}(N-2) + 2N\right)\left(1 + \frac{(N-2)}{N}\sqrt{M}\right). \tag{6.9}$$

The scaling of K'_S is portrayed in Fig. 6.11. As can be seen, increasing M does have a notable impact on the FSL losses. Moreover, the absolute increase is relatively constant regardless of M. Comparing these results with Fig. 6.10, it is clear that after a certain point, increasing M will add more losses in the switches than are reduced in the capacitors. What exactly this optimum is, however, depends heavily on the converter's iVCR and power-density, and a number of technology parameters.

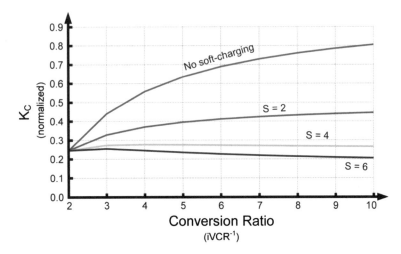

Fig. 6.10 Scaling of a N:1 Dickson converter's K_C versus N/$iVCR^{-1}$ without soft-charging, and with soft-charging using a selection of soft-charging factors, S

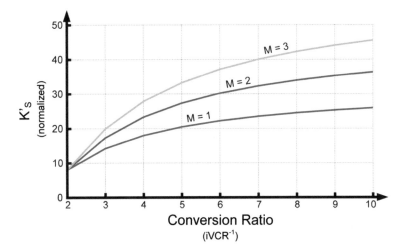

Fig. 6.11 Scaling of a N:1 Dickson converter's K'_S versus N/$iVCR^{-1}$ without MSC, and with MSC using a selection of factors, M

6.2.5   Other Topologies

To verify that SO and MSC can be applied to any topology, their use is demonstrated in the Series-Parallel- and Fibonacci converter families in this Section.

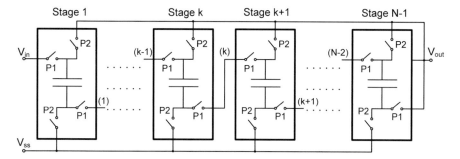

Fig. 6.12 A regular N:1 Series-Parallel SC converter with four switches per stage

Series-Parallel Converter

An N:1 Series-Parallel converter first connects all of its flying capacitors in series between V_{in} and V_{out}, and then puts them in parallel between V_{out} and V_{ss}, as shown in Fig. 6.12 [See09]. The latter phase is thus always hard-charged for all flying capacitors, which implies that the relative gain of applying any soft-charging method can never be greater than a factor of 2. The former phase, in contrast, can be soft-charged. Although, this does require a modification of the switch structure of the Series-Parallel stages. Normally, each stage in a Series-Parallel converter has three switches, one of which is used to connect to the top-side of the next stage. To enable stage outphasing, this switch must be split into two to create an intermediate node, (k). Unlike with the Dickson converter, this is not a common operation because this switch only needs to block approximately V_{out}. Furthermore, because with the application of stage outphasing the intermediate nodes (k) are DC, where the DC level is the voltage at which the adjacent stages exchange charge, the voltage these switches need to block is actually increased to an integer multiple of V_{out} depending on the stage. In short, the required modifications to the series-parallel converter significantly worsen its switch utilization, which means it is not a good candidate for SO/MSC at high power-densities.

The Series-Parallel converter with stage outphasing is portrayed in Fig. 6.13. As can be seen, once the intermediate nodes are present, the key to achieving stage outphasing is simply to split every stage into two cells that run in anti-phase, and introduce a phase shift between adjacent stages. The charging state is consequently split into multiple steps. The number of steps is not limited to two, but can go up to the number of cascaded stages in the full converter: $N-1$. This is due to fact that all stages are put in series in the charging state, while for the Dickson converter, there are only 2 stages interacting at any given time. Although not shown in Fig. 6.13, to allow multiphase soft-charging, the necessary step is to add extra intermediate nodes, $(k,1)$, $(k,2)$, ... just like was done for the Dickson converter. This way, soft-charging can be achieved in a scalable way, at the cost of extra switches. With S the number of soft-charging steps, the topology's capacitor utilization can be deduced:

Fig. 6.13 Application of
Stage Outphasing to a
Series-Parallel converter

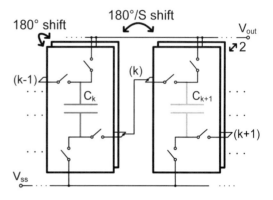

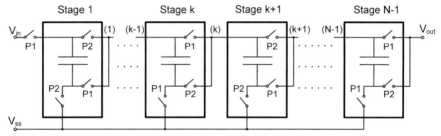

Fig. 6.14 A regular Fibonacci SC converter with N-1 stages

$$K_{C,series-parallel,sc} = \left(\frac{N-1}{N}\right)^2 \frac{S+1}{2S}. \tag{6.10}$$

Fibonacci Converter

The Fibonacci converter [UIOH91], shown in Fig. 6.14 is more suited to advanced
multiphasing compared with the Series-Parallel topology. Although, as illustrated in
Fig. 6.15, it does need some modifications. That being said, as was the case with the
Dickson converter, the changes that enable SO appear naturally when the converter's
output voltage is approximately V_{out}, which means that the switch utilization is
unaltered. In addition, only the discharging state of the final stage is hard-charged,
though this is normally the largest stage in terms of capacitor size. Furthermore,
similar to the Series-Parallel topology, all stages can be outphased relative to each
other, thus enabling larger efficiency gains.

Applying stage outphasing to the Fibonacci converter, however, does have the
downside that the soft-charging itself is not well controlled: Contrary to the Dickson
converter, where there is always a charge-exchange between a single charging-
and discharging stage, the Fibonacci has multiple charging- and discharging stages
interacting with each other. When stage outphasing is used, there are consequently
discharging cores that will be charging other discharging cores, thus negatively

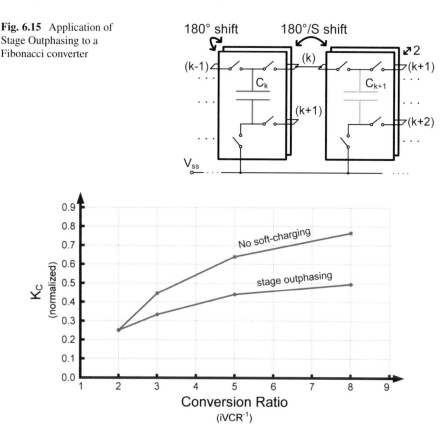

Fig. 6.15 Application of Stage Outphasing to a Fibonacci converter

Fig. 6.16 Scaling of a Fibonacci converter's K_C versus $iVCR^{-1}$ without soft-charging, and with stage outphasing

impacting the converter's output impedance. Nonetheless, as illustrated by Fig. 6.16, there is still a substantial net-gain.

6.3 Implementation

6.3.1 System

The proposed techniques of Stage Outphasing and Multiphase Soft-Charging are implemented in a 3:1 Dickson converter with $M = 2$. Using (6.8) one can calculate that this leads to a 60% higher effective capacitance density. Because a 3:1 converter only has two stages, each stage has a soft-charged- and a hard-charged state. The hard-charged state, however, does not require the same amount of time to complete as the soft-charged state, which is spread over multiple phases. Therefore, the

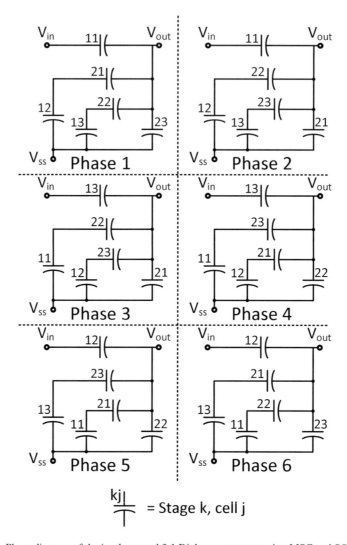

Fig. 6.17 Phase diagram of the implemented 3:1 Dickson converter using MSC and SO

number of phases of the hard-charged state is reduced from four to two in this implementation. This has the advantage of also reducing the total number of phases and cells per stage to six and three, respectively, which in turn reduces frequency and overhead. The resulting phase diagram of the implemented converter is shown in Fig. 6.17.

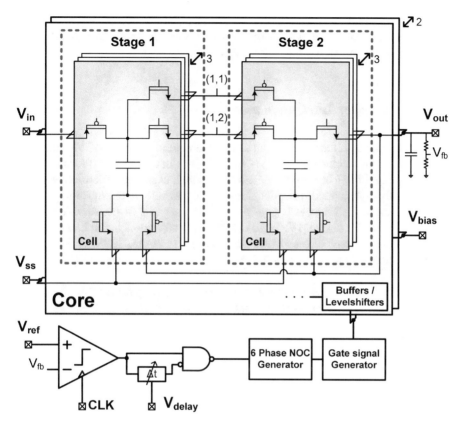

Fig. 6.18 System overview of the 3:1 converter, showing the controller and transistor-level implementation of the converter cores

Figure 6.18 gives an overview of the full converter, including the controller. Because of SO, only half of the phases are used in each stage. A second core can consequently be added that runs in anti-phase relative to the first, and uses the other half of the phases. As a result, the amount of charge that is transferred to the output each switch event is equalized, and the output ripple is subsequently reduced. In addition, a small amount of decoupling capacitance equal to approximately 9% of the total flying capacitance is integrated for the output terminal.

The hysteretic controller runs at a clock frequency of 1.6 GHz, which means that the converter itself runs at 267 MHz. A rising edge-detect circuit with a tunable delay cell generates a pulse with a variable width which is in turn used to clock a 6 phase ring-connected one-hot-coded oscillator based on SR flip flops. The width of the pulse ultimately determines the width of the dead time between every phase. The six NOC phases are then used to generate the required gate signals. Due to the relatively small amount of unique gate signals, this is done centrally. In fact, only the capacitive level shifters [MPK+15] and buffers are placed close to each cell.

Even though the intermediate nodes, (1,1) and (1,2), are not DC because of SO, they are used to power the top-side buffers and power transistors with little effect on the soft-charging operation of the converter.

6.3.2 Capacitor Implementation

The flying capacitors are implemented using a combination of MOM- and MOS capacitors, with thick-oxide devices used in the first stage due to their higher voltage rating. MOS capacitors, while having a high capacitance density, also have a notoriously high parasitic coupling due to their proximity to the substrate. To limit the influence of the latter, it is possible to bias the wells of the MOS capacitors with a suitable bias circuit [VPWRB11]. Multiple converters in the literature have implemented such techniques [MPK$^+$15, JLH$^+$15, LCSA13, SPSS15]. Figure 6.19 shows a schematic comparison of a selection of these to the implementation used in this work.

The accumulation-based capacitor of [JLH$^+$15] (Fig. 6.19a) has the advantage of having a smaller parasitic junction capacitor, C_{pwdnw}, over the P-well to deep N-well (DNW) junction, compared to the channel capacitor, C_{chan}, of the inversion-based capacitor used in [LCSA13] (Fig. 6.19b). This is due to the fact that the latter has higher doping concentrations. The implementation of [LCSA13], on the other hand, does bias the N-well using a sufficiently large resistor, causing the channel capacitor to effectively be placed in series with the N-well to substrate junction capacitor, C_{nwpsub} and leading to significant reduction in parasitics.

The proposed implementation combines the advantages of both to minimize the effective parasitic coupling as much as possible. The result is an accumulation-based capacitor that uses a dedicated high voltage bias, V_{bias}, to bias the DNW. Furthermore, by using two front-to-front diode-connected PMOS devices, the high-impedance is realized using at least 100 times less area overhead than the approach in [LCSA13]. Figure 6.20 compares the parasitic coupling of the used implementation to the state of the art. As can be seen, the used MOS capacitors have significantly lower coupling. For a bias voltage equal to the converter's input voltage, an improvement of 18% is witnessed, while higher relative bias voltages can push this improvement to 30% and more. Overall, thanks to the proposed implementation, the ratio of parasitic- to flying capacitance is made smaller than 1%.

Note also that all of the shown capacitor implementations have their parasitic coupling on the flying capacitor's top plate. This has the inherent advantage that the parasitic capacitor acts as an extra parallel 1:1 converter which supplies power to the load, rather than an additional load which needs to be supplied by the SC converter itself if the parasitic coupling were present on the bottom plate [MBS13].

Fig. 6.19 Schematic representation of the biasing techniques discussed in (**a**) [JLH+15], (**b**) [LCSA13], and (**c**) the one used in this work

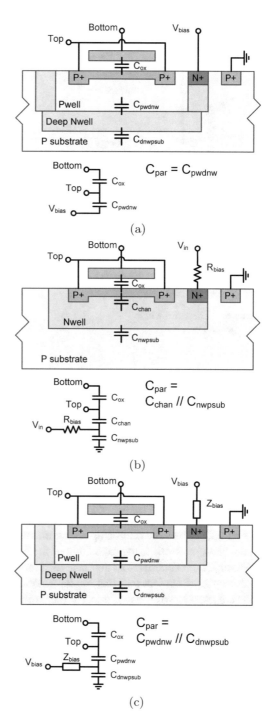

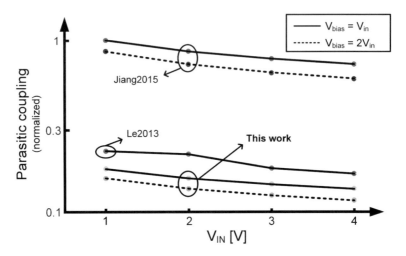

Fig. 6.20 Comparison of the simulated parasitic coupling of the used implementation to literature [JLH⁺15, LCSA13]

Table 6.1 Transistor width sizes per cell in μm

		PMOS	NMOS
Stage 1	Top-side	250	100
	Bottom-side	279	167
Stage 2	Top-side	120	180
	Bottom-side	197	236

6.4 Experimental Verification

6.4.1 Realization

The design is realized in a 28 nm baseline CMOS process using a total flying capacitance of 1.5 nF and with the power transistor widths as given in Table 6.1. Figure 6.21 shows the microchip which measures $0.260 \times 0.448\,\mu m^2$. As discussed in the previous section, the capacitors of the first stage are partly implemented using thick-oxide MOS capacitors and take up a larger area as a result.

Because of the fact that this implementation uses a total of six phases of which four are used to soft-charge either the charging- or discharging state, every cell of stage one will connect to a certain cell of the second stage two times instead of just once over a full cycle. To minimize the resistive losses in the interconnects between cells and stages, these pairs of cells are positioned closest to each other.

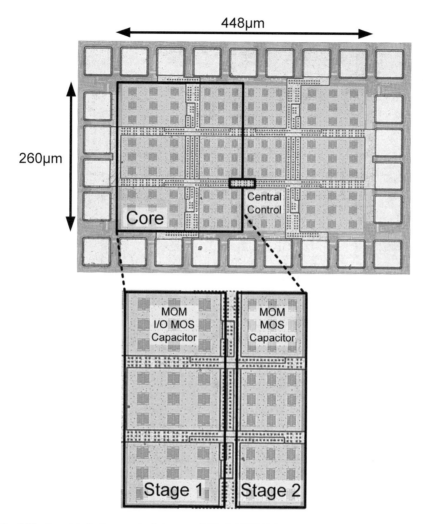

Fig. 6.21 Annotated micrograph of the monolithic 3:1 SC Dickson converter using SO and MSC, measuring $0.117\,\text{mm}^2$ without bond pads

6.4.2 Measurement Setup

The measurement setup, illustrated in Fig. 6.22, is largely the same as the one used in Sect. 4.5.1. The main difference is that the Rohde & Schwarz SMBV100A vector signal generator supplies the frequency reference rather than the Rigol DG3101A, due to its higher frequency limit [Roh17, RIG08]. Furthermore, an extra Keysight 34401A multimeter is added to measure the voltages on the intermediate nodes [Key12].

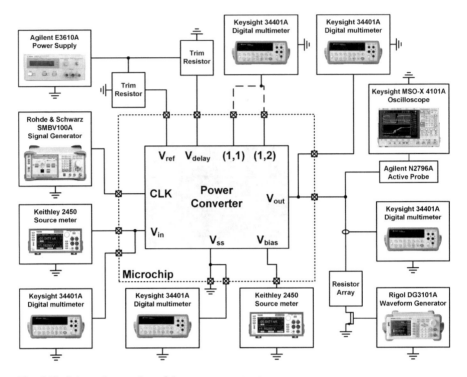

Fig. 6.22 Schematic overview of the measurement setup

The generation of V_{bias} is not realized on-chip in this design. For all below measurements a voltage of 8.5 V was supplied externally using a Keithley 2450 source meter [Kei11], drawing less than 60 nA at an ambient temperature of approximately 295 K. While this voltage is much higher than the technology's nominal voltage, it is still well below the junction's breakdown voltage [CR16, ILPY17].

6.4.3 Measurements

Intermediate Nodes

To verify the working principle of MSC, the converter is measured under open-loop operation with increasing load currents. The results hereof are demonstrated in Fig. 6.23. The theoretical prediction is done using a charge-analysis of the topology and states that, given V_{in} and V_{out}, $V_{(1,1)}$ and $V_{(1,2)}$ can be determined as

$$V_{(1,1)} = \frac{2}{3} V_{in} \tag{6.11}$$

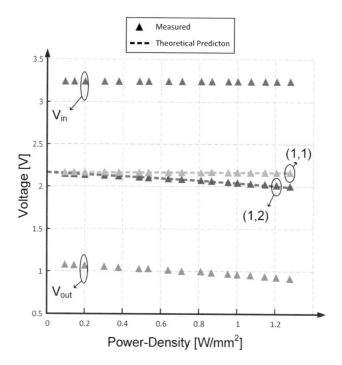

Fig. 6.23 Comparison of measured DC voltages of intermediate nodes, (1,1) and (1,2), under open-loop operation with $f_{clk} = 1.3$ GHz and $V_{in} = 3.25$ V, to theoretical predicted values based on V_{in} and V_{out} measurements

$$V_{(1,2)} = \frac{1}{3}V_{in} + V_{out}, \tag{6.12}$$

which can be rewritten to

$$V_{(1,1)} - \frac{V_{in} + V_{out}}{2} = \frac{V_{in} + V_{out}}{2} - V_{(1,2)}. \tag{6.13}$$

As can be seen from Fig. 6.23, the intermediate nodes are pulled apart in order to supply more charge each clock cycle. At the same time, in agreement to (6.13), they remain symmetrical with respect to the average of V_{in} and V_{out}, which implies that the charge transfers of both MSC phases are approximately equal. Overall, the measured behavior of the intermediate nodes matches the theoretical prediction of (6.11) and (6.12) with less than 1% error.

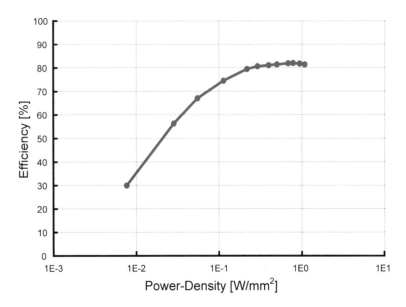

Fig. 6.24 Measured full-system closed-loop efficiency versus output power-density for $f_{clk} = $ 1.6 GHz, $V_{in} = 3.2$ V and $V_{out} = 0.95$ V

Efficiency

The converter's closed-loop efficiency is shown in Fig. 6.24, which is measured using Kelvin contacts and includes all system losses. A peak efficiency of 82% is achieved at power-densities of 0.65–1.1 W/mm^2 or output powers of 76–126 mW. Thanks to the hysteretic controller, the input-referred quiescent current of the converter is approximately 0.9 mA, leading to a wide efficient output power range.

Controller

Figure 6.25 demonstrates the operation of the hysteretic controller as it is tested under worst-case load-regulation conditions. For this measurement, the load power is switched from the maximum load power of 126 mW to zero and back with a transient time of 18 ns.

6.4.4   Comparison

The presented work is compared to the state of the art of fully integrated SC DC-DC converters in Fig. 6.26 and Table 6.2. Thanks to the presented soft-charging

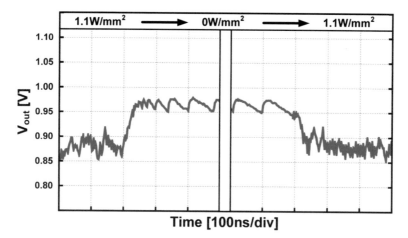

Fig. 6.25 Measured worst-case load-step transient response with $f_{clk} = 1.6\,\text{GHz}$, $V_{in} = 3.2\,\text{V}$ and $V_{ref} = 0.95\,\text{V}$

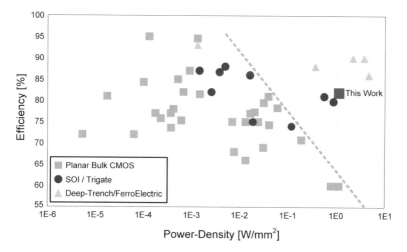

Fig. 6.26 Extensive efficiency versus power-density comparison of the presented converter to the state of the art of fully integrated SC converters, based on [SBM$^+$]. The dotted line represents the current state of the art for planar bulk CMOS designs

techniques and the capacitor implementation, the realized converter achieves a significantly higher efficiency-power-density combination than any other published planar bulk SC converter, and also outperforms designs using SOI or TriGate technologies. Relative to the current state of the art of planar bulk CMOS SC converters, the presented work has either 29× higher power-density for the same efficiency, or 3× lower losses for the same power-density.

Table 6.2 Comparison to state of the art

	This work [BS17b, BS17a]	[SSP+09]	[MPK+15]	[LSA11]
Technology	28 nm	32 nm	90 nm	32 nm SOI
Capacitors	MOM + MOS	MOM	MOS	MOM + MOS
$iVCRs$	3:1	1:2	2:1	3:2 2:1 3:1
V_{in} [V]	3.2	1	2.4–2.6	2
V_{out} [V]	0.95	1.5	0.9–1.3[a]	0.5–1.15
P_D [W/mm^2]	1.1	1.1	0.77	0.86
η @ P_D	82%	60%	60%	79.8%
Closed loop?	Yes	Yes	No	No
Area [mm^2]	0.117	0.067	2.14	0.378
$iVCR$ @ P_D	3:1	1:2	2:1	2:1
P_N @ P_D	22%	67%	67%	25%
PDN FoM[b]	34.5%	N/A	183.2%	56.0%
High density FoM[b] $[(\mathrm{W/mm^2})^{-\frac{1}{3}}]$	0.37	1.64	1.83	0.67

[a]Estimate based on graphs
[b]See (2.67) and (2.75), smaller is better

While efficiency versus power-density comparisons such as the one portrayed in Fig. 6.26 are useful, they do not show the full picture. In particular, no information regarding the voltage conversion ratio is included, even though this is a crucial factor for many applications, including the reduction of PDN losses discussed in Sect. 6.1. Therefore, Fig. 6.27 compares the presented work to the state of the art using the PDN- and HD FoM derived in Sect. 2.4. As can be seen, the presented work performance significantly better compared to both planar bulk and SOI/Trigate designs, thanks to its high efficiency, power-density, and voltage conversion ratio. In fact, the presented work achieves results that are nearly comparable to designs using expensive Deep-Trench capacitors, which have more than 20 times the capacitance density and are far from common in today's technology nodes.

6.5 Conclusion

In this chapter, the need for combining high efficiency, high VCR, and high power-density in fully integrated DC-DC converters was discussed, together with the difficulty of achieving this due to the limited capacitance density available in modern technology processes. Two techniques, called Stage Outphasing and Multiphase Soft-Charging, were introduced that make use of the advanced multiphasing concept to soft-charge charge transfers between flying capacitors. As such the effective flying capacitance can be increased.

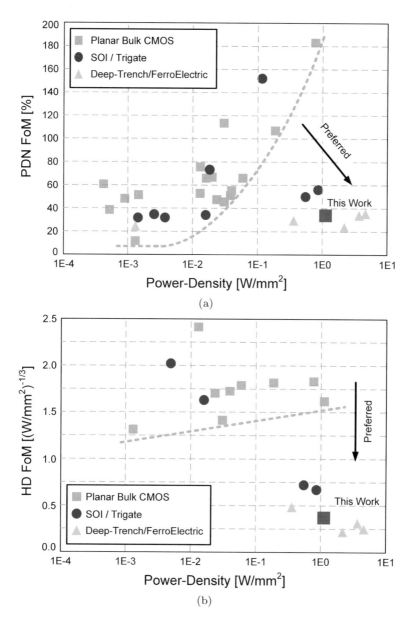

Fig. 6.27 Comparison of the presented work to the state of the art using (**a**) the power delivery network- and (**b**) the high-density figure-of-merit. The dotted line represents the state of the art for planar bulk CMOS designs

The techniques were analyzed for a Dickson SC converter and their impact on the topological parameters was discussed. Furthermore, this chapter also explored their use in other switched-capacitor topologies. A 3:1 Dickson SC converter was realized that implements these techniques to achieve a state of the art $1.1 \, \text{W/mm}^2$ power-density and 82% efficiency combination using common capacitor technologies.

Chapter 7
Continuously-Scalable Conversion Ratio Topologies

So far in this book, multiple techniques have been proposed to improve the performance of switched-capacitor converters at both low- and high power-densities. In essence, all of these techniques achieve purely capacitive soft-charging, enabled by the concepts of advanced multiphasing. This chapter will extend on the same concepts to challenge the very core design principles of switched-capacitor converters, and, in the process, open up a fundamentally different dimension of the SC converter design space.

7.1 Background and Motivation

Since their first usage in monolithic systems, switched-capacitor DC–DC converters' basic principles have remained largely unchanged [Dic76]. Because capacitors inherently incur charge-sharing losses when switched between different voltages by means of a battery or other capacitors, the only viable strategy to make efficient switched-capacitor converters has been to limit the voltage swing across the capacitor terminals as much as possible. This can be explained as follows. From (2.3), it can be appreciated that the charge-sharing loss of a charge transfer is proportionate to the capacitor's voltage swing squared, ΔV^2. At the same time, the transferred charge, q, is proportionate to ΔV. Combining both, the "efficiency" of a charge transfer can be established:

$$\frac{q}{E_{cs}} \propto \frac{1}{\Delta V}. \tag{7.1}$$

In other words, if a charge transfer is to be efficient, the voltage swing needs to be minimized. Consequently, SC topologies with approximately fixed capacitor voltages are used, which only able to supply a specific rational voltage conversion ratio efficiently, as shown in Fig. 7.1. When a lower voltage is required than dictated

© Springer Nature Switzerland AG 2020
N. Butzen, M. Steyaert, *Advanced Multiphasing Switched-Capacitor DC-DC Converters*, https://doi.org/10.1007/978-3-030-38735-8_7

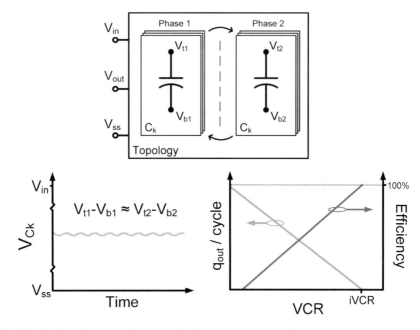

Fig. 7.1 High level concept of a conventional two-phase switched-capacitor converter

Fig. 7.2 Output charge per cycle and efficiency of a gearbox switched-capacitor converter

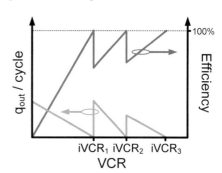

by the topology's iVCR, or when a larger output power is needed, the converter can only obey by increasing the capacitors' voltage swing and thereby lowering its efficiency. As such, there is a very strong connection between a conventional SC converter's VCR, efficiency, and output power. Especially for applications that require a large output voltage range, like DVFS-enabled digital processors, this has proven to be a problem.

In the literature, the usual solution is to combine multiple topologies together in a gearbox converter [EDBC13, JSB16, SM14b], which results in the typical saw-tooth curve portrayed in Fig. 7.2. Such converters do, however, have multiple disadvantages. For one, to maintain a high efficiency over a wide voltage range, many topologies are usually required, each adding extra transistors which often end

Table 7.1 Peak efficiency of many-ratio monolithic gearbox converters in the literature, with- and without topology control

Publication	With topology control	Without topology control
[SM14b]	85%	
[JLMM18]		84%
[JSB16]		95%
[BWG+13]	72%	
[LYM+17]	87%	94%

up as extra series resistances in other topologies. Moreover, the required control overhead to switch between different topologies at the right moment increases rapidly with the number of topologies, and thus induces significant losses. To demonstrate this, a comparative overview of the state of the art of gearbox converter with 15 iVCRs or more is given in Table 7.1. In general, the gearbox converters that do not implement topology control have a substantially higher peak efficiency compared with those that do. The best example is the design in [LYM+17] where the efficiency reportedly drops 7% by activating the control loop, which corresponds to more than a doubling of the normalized losses. In addition, it was pointed out in [SRS18] that, as monolithic switched-capacitor converter designs implement more ratio's, they also suffer in terms of power-density.

Previous chapters demonstrated that charge transfers between capacitors can be soft-charged using the advanced multiphasing concept, by spreading the transfer of charge itself out over multiple steps or phases. Consider a capacitor that was soft-charged this way using S phases, its charge transfer "efficiency" can easily be deduced:

$$\frac{q}{E_{cs}} \propto \frac{S}{\Delta V}. \tag{7.2}$$

Thus, with advanced multiphasing, there are two, rather than one, design parameters. This means that minimizing the capacitors' voltage swing is no longer required to obtain an efficient converter. After all, the efficiency of the converter can instead be guaranteed using the soft-charging parameter, S. Therefore, advanced multiphasing offers the exciting opportunity to make switched-capacitor converters with large capacitor voltage swings.

This chapter has the following outline: Sect. 7.2 proposes a new type of switched-capacitor topology that maintains high efficiency despite large capacitor voltage swings. Because this type of converter is radically different than conventional SC converters, a closer look will be taken to how it can be modeled and optimized in Sect. 7.3. Section 7.4 discusses an example implementation of such a topology, while Sect. 7.5 verifies its working principle with measurement results. Finally, Sect. 7.6 gives a brief summary and discusses the important conclusions of this chapter.

7.2 Working Principle

Figure 7.3 illustrates the high level concept of the proposed topology. As can be seen, the basic idea is to maximize the voltage variation of the flying capacitor(s) by making it swing between approximately V_{in} and 0 V and back, regardless of the output voltage. The firm relation between output charge per cycle (or efficiency) and VCR is consequently broken. As mentioned above, the addition of the soft-charging phases boosts the efficiency of the converter. Another way to think about this concept relative to that of a conventional SC converter is that conventional topologies have 100% efficiency to begin with, but no output power, causing most of the design of such a converter to revolve around getting the required power out of it (Fig. 7.1). The proposed topology, in contrast, has large output power, but no efficiency, and this is precisely why soft-charging is needed.

Opening up the capacitors' voltages to have no fixed bias voltage, dramatically increases the number of topologies one can make, especially when also using a lot of phases. Therefore, to minimize the design complexity, only single-capacitor topologies are considered in this work. Note that this implies that said capacitor will be soft-charging with phase-shifted versions of itself. In order to simplify a practical implementation, it is assumed that the capacitor polarity stays the same over a full cycle. Furthermore, the main goal here is to make a step-down converter.

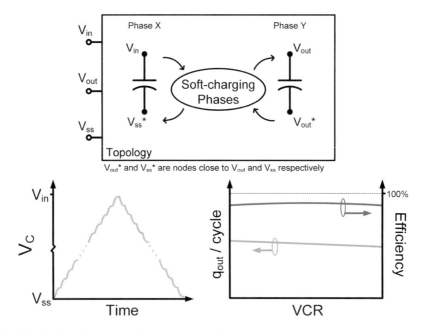

Fig. 7.3 High level concept of the proposed switched-capacitor converter

7.2.1 Useful Soft-Charging

To boost the efficiency, soft-charging must be applied carefully such that each charge transfer is useful to the full conversion effort. For example, in Chap. 6 each soft-charging step transfers charge from the ground/V_{ss} terminal to the output, resulting in an increase of the output charge per cycle and subsequently the system efficiency. As such, this type of soft-charging is definitely useful in this context. On the other hand, if the SPCR technique of Chap. 4 would be applied to flying capacitors that are topologically speaking the same, rather than parasitic capacitors, this would not be the case. After all, the capacitors would simply exchange charge with themselves, and not to/from the converter terminals. With these examples in mind, a necessary condition for a charge transfer to be considered useful, at least when there is only a single-capacitor topologically speaking, is that it results in charge being added/subtracted to the converter terminals. Also, these charge exchanges with the terminals should be in line with the desired voltage conversion ratio: in step-down converters, for example, charge should be subtracted from the V_{ss} terminal.

Translating said condition to the phases of the converter, this implies that the capacitor should always have one terminal connected to one of the converter terminals. Moreover, during a set of soft-charge transfers, the capacitor should stay connected to the same terminal, thereby gradually (dis)charging the converter terminal.

7.2.2 Topology Structure

While soft-charging is an essential element of the topology, its structure is actually defined by the phases that would be there even without soft-charging. In some sense, the topology is built on these phases, and as such we refer to them as the *cornerstone phases* of the topology. The definition above means that, during a cornerstone phase, each capacitor terminal must be connected to one of the converter terminals. In addition, to satisfy the condition that the capacitor has to stay connected to the same terminal during a set of soft-charge transfers, two consecutive cornerstone phases should always share a capacitor terminal connection.

From the basic concept description above, it was established that the voltage swing of the converter should be as large as possible. For a step-down converter, the largest achievable voltage swing where the polarity of the capacitor does not change equals V_{in}. As such, two cornerstone phases are deduced: One where the capacitor is connected to V_{in} and V_{ss}, and one where the capacitor voltage is zero. In the latter cornerstone phase, the capacitor could technically be connected to any of the three converter terminals. However, if it were to connect to the input- or ground terminal in this phase, this will inevitably lead to charge being pumped to said terminals, which is of course to be avoided for step-down conversion. This can be explained as follows: The input terminal is assumed to be the highest voltage in the converter. Therefore, given that the polarity of the capacitor does not change,

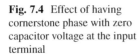

Fig. 7.4 Effect of having cornerstone phase with zero capacitor voltage at the input terminal

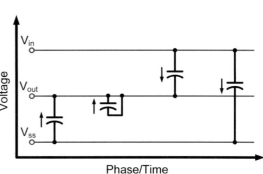

Fig. 7.5 Cornerstone phases of presented topology

the capacitor cannot have its bottom plate connect to V_{in} without having its top-plate connect to V_{in} as well. This means that, instead, the cornerstone phase before and after this cornerstone phase must both have the top-plate connected to the input while the bottom-plate connects to a lower voltage terminal, thus leading to charge being transferred to the input as illustrated in Fig. 7.4. A similar reasoning can be applied to the ground terminal.

Because the two cornerstone phases so far do not share a capacitor terminal connection, at least two more cornerstone phases need to be added in between. There are two valid phases remaining that match the earlier made assumption of capacitor polarity. In one phase, the capacitor connects to V_{in} and V_{out}, while in the other it connects to V_{out} and V_{ss}. There is only way of combining these with the previous cornerstones, such that charge is transferred to and from the right terminals. This set of cornerstone phases is shown in Fig. 7.5.

It can be appreciated that the top plate of the capacitor alternates between the input- and output terminals, while the bottom plate alternates between V_{out} and V_{ss}. This is not only advantageous in terms of the number of switches (or switch area) that will be required, but also simplifies the implementation of soft-charging: Because each capacitor terminal goes through a separate charge- and discharge cycle, soft-charging is simply enabled by adding intermediate nodes for the top- and bottom plate separately. As was the case with the charge redistribution buses and the nodes of the MSC technique of Chaps. 4 and 6, respectively, these intermediate nodes will spread evenly between the boundary conditions of V_{in}/V_{out} and V_{out}/V_{ss} for the respective capacitor plate.

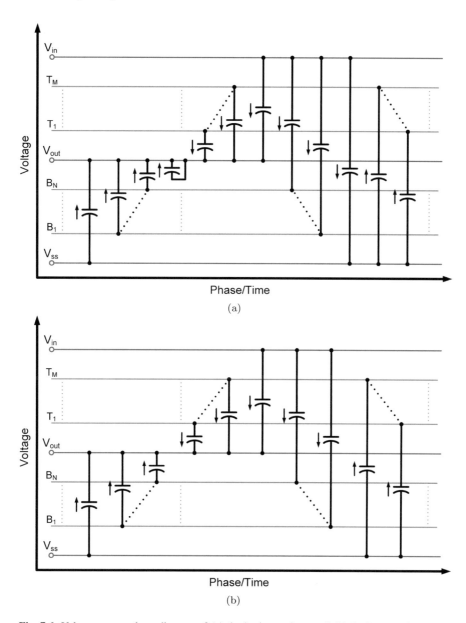

Fig. 7.6 Voltage versus phase diagram of (**a**) the basic topology and (**b**) the improved topology for step-down conversion

Figure 7.6a shows the topology with M soft-charging phases at the top-side (nodes T_1 to T_M) and N soft-charging phases at the bottom-side, with corresponding nodes B_1 to B_N. At this point, all of the charge-transfers are useful for step-down conversion, except for those of the two cornerstone phases we started off with.

These phases, while important in defining the topology structure, can thus better be omitted, leading to the improved topology of Fig. 7.6b. Here, the top-side nodes spread evenly between $V_{in} - \Delta V_{botside}$ and $V_{out} + \Delta V_{botside}$ with $\Delta V_{botside}$ the voltage difference between adjacent bottom-side nodes, rather than V_{in} and V_{out} because of the change in boundary conditions. In addition, removing said phases is a key step in enabling outphasing, which is discussed in Sect. 7.2.3. All subsequent analyses will use the improved topology.

7.2.3 Outphasing

The improved topology has the additional benefit of being compatible with Outphasing, introduced in Chap. 6. Instead of outphasing between different stages, though, here it is possible to introduce a phase shift between the charge transfers of a single stage. This is illustrated in Fig. 7.7. As can be seen, all of the connections that take

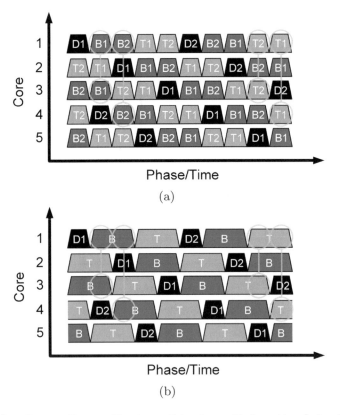

Fig. 7.7 Phase diagram of improved topology with two top- and bottom-side soft-charging phases. (**a**) Is without- (**b**) is with outphasing. D1 and D2 are the cornerstone phases, and B/B1/B2 and T/T1/T2 are the bottom- and top-side intermediate nodes, respectively. A selection of connections between core 1 and cores 3/4 is highlighted

place through the intermediate nodes B1/B2/T1/T2 between the different converter cores can be replicated with half as many intermediate nodes if the soft-charging phase durations are doubled in length relative to the cornerstone phase durations. In other words, the cornerstone phases introduce a phase shift that the soft-charging phases benefit from. The end result is that the number of intermediate nodes, as well as the number of switches that connect the capacitor to those nodes, is halved. Needless to say, this is a substantial improvement in terms of switch losses and overhead area.

7.3 Modeling

The presented topology has a fundamentally different mode of operation compared with any other switched-capacitor DC–DC converter in the literature. Hence, the output impedance model described in Chap. 2 does not necessarily apply. In this section, an equivalent model of the introduced topology is derived, and a closer look is taken to the trade-offs in the design of this type of converter.

7.3.1 Intrinsic Behavior

For the following analysis it is assumed that the converter is in steady-state and that all charge transfers are completed fully. In addition, V_{ss} is presumed to be zero, while the output has infinite decoupling. With these assumptions in mind, two sets of equations can be derived by applying a charge-analysis to each intermediate node of Fig. 7.6b:

$$\begin{cases} V_{B,1} & = V_{B,2} - V_{B,1}, \\ V_{B,i+1} - V_{B,i} & = V_{B,i} - V_{B,i-1} \quad \forall i \in \{2, \dots, N-1\}, \\ V_{out} - V_{B,N} & = V_{B,N} - V_{B,N-1}, \end{cases} \quad (7.3)$$

$$\begin{cases} V_{T,1} - 2V_{out} + V_{B,N} & = V_{T,2} - V_{T,1}, \\ V_{T,j+1} - V_{T,j} & = V_{T,j} - V_{T,j-1} \quad \forall j \in \{2, \dots, M-1\}, \\ V_{in} - V_{T,M} - V_{B,1} & = V_{T,M} - V_{T,M-1}, \end{cases} \quad (7.4)$$

where $V_{B,i}$ and $V_{T,j}$ are the bottom- and top-side nodes, respectively. Moreover, upon visual inspection, the output- and input charge per cycle, q_{out} and q_{in}, are determined to be

$$\frac{q_{in}}{C_{tot}} = (V_{in} - V_{T,M}) + (V_{out} - V_{B,1}), \quad (7.5)$$

$$\frac{q_{out}}{C_{tot}} = (V_{T,1} - V_{out}) + V_{B,N} + (V_{in} - V_{T,1}) + (V_{T,1} - 2V_{out} + V_{B,N}). \quad (7.6)$$

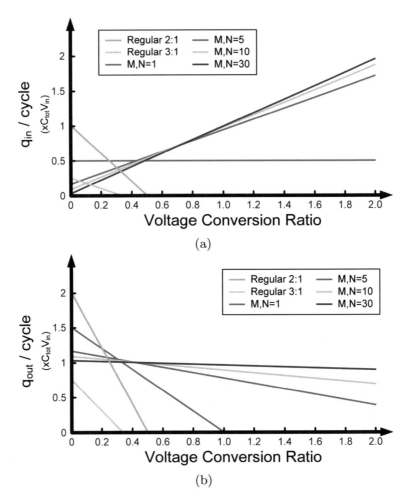

Fig. 7.8 Input- (**a**) and output- (**b**), charge per cycle of presented topology for different number of top-and bottom-side soft-charging phases, compared with a regular 2:1 and 3:1 SC converter

Combining this with (7.3) and (7.4) then yields

$$\frac{q_{out}}{C_{tot}} = \frac{M+2}{M+1}V_{in} - \frac{M+N+4}{(M+1)(N+1)}V_{out}, \qquad (7.7)$$

$$\frac{q_{in}}{C_{tot}} = \frac{1}{M+1}V_{in} + \frac{MN+M-2}{(M+1)(N+1)}V_{out}, \qquad (7.8)$$

Figure 7.8 plots these formula's for increasing values of M and N, and also compares them to those of a regular 2:1 and 3:1 SC converter. The first thing to point out is that the presented topology indeed has a very different behavior compared

Fig. 7.9 Gyrator model of
presented topology

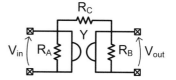

with the regular converters. For one, q_{in} and q_{out} are not proportionate to each other, and there is no sign of an ideal VCR. Instead, as M and N increase, the output charge has a much more gradual slope and even converges to being constant regardless of the VCR and including step-up conversion, as if the converter were a current source. Furthermore, the level it converges to is high compared with the regular converters output charge, at least for the VCR range where the latter are still efficient. The input charge, on the other hand, is proportionate to the VCR, which leads to a high efficiency.

For M and N infinitely high, the behavior described above matches that of a gyrator. A gyrator is a lossless two-port element where the current through a port only depends on the voltage of the other port, and is characterized by its (trans)conductance, Y.

$$\begin{bmatrix} I_{in} \\ I_{out} \end{bmatrix} = \begin{bmatrix} 0 & Y \\ Y & 0 \end{bmatrix} \begin{bmatrix} V_{in} \\ V_{out.} \end{bmatrix}$$ (7.9)

Figure 7.9 shows the equivalent gyrator model of the presented topology, with additional resistance parameters, R_A, R_B, and R_C. Using (7.7) and (7.8), these parameters can be deduced:

$$Y = f_{sw}C_{tot}\frac{MN + N + M}{(M + 1)(N + 1)}$$ (7.10)

$$R_A = \frac{-1}{f_{sw}C_{tot}}(M + 1)(N + 1)$$ (7.11)

$$R_B = \frac{1}{f_{sw}C_{tot}}\frac{(M + 1)(N + 1)}{M + 2}$$ (7.12)

$$R_C = \frac{1}{f_{sw}C_{tot}}\frac{(M + 1)(N + 1)}{N + 2}.$$ (7.13)

From (7.10) it can be appreciated that, despite the fact that the converter behaves like a current source for constant V_{in}, the output current is also proportionate to the switching frequency. Frequency control schemas like the lower-bound hysteretic control commonly used in SC DC–DC conversion are thus compatible with this topology. Moreover, even if there is no dedicated output capacitor, this control can be done in the voltage domain because the capacitance of the converter itself behaves like a decoupling capacitor as seen from the output.

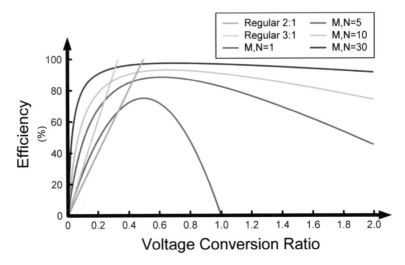

Fig. 7.10 Efficiency of presented topology for different number of top- and bottom-side soft-charging phases, compared with a regular 2:1 and 3:1 SC converter

Figure 7.10 plots the efficiency of the presented topology for different values of *M* and *N*. As can be seen, the topology obtains a wide efficient voltage conversion ratio range, which is substantially larger than that of a single regular SC converter. In addition, the larger the number of top- and bottom-side soft-charging phases, the higher the obtained efficiency. Naturally, *M* and *N* do not have to be the same value. For VCRs smaller than 2:1, the losses at the top-side are more significant relatively speaking, which makes M more important. Similarly, N gains in importance for VCRs larger than 2:1, including step-up conversion.

7.3.2   Finite Switch Resistance

In a practical implementation, the finite switch resistance prohibits the full completion of charge transfers. With regular SC converters, the effect of the finite switch resistance can be approximated by looking at the converter in the worst-case asymptotic limit where the capacitors behave like perfect DC voltage sources, and the currents going through the switches are also perfect DC (see Chap. 2). In the presented converter, however, this limit has little meaning because it is precisely the highly variable capacitor voltage which is key to the converter operation.

At the same time, there does not appear to be another way for the effect of the switch resistance to be incorporated in a manageable closed-form solution or approximation. Unfortunately, this forces us to work with numerical methods or simulations. Figure 7.11 shows the result of such numerical calculations for an example converter and with varying total switch conductance. In the reference case, the normalized losses are approximately 5.6% for a VCR of 2:1, with the charge

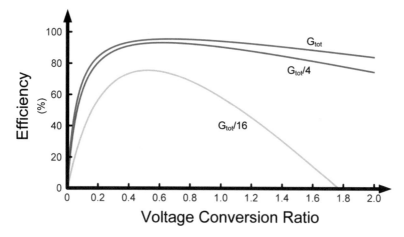

Fig. 7.11 Efficiency of presented topology with M, $N = 16$, and scaling total switch conductance, G_{tot}

transfers almost fully completing (99.93%). Reducing the switch conductance by a factor of 4 and 16, corresponding to respective charge transfer completions of 81 and 30%, these losses increase to 8.1 and 35.5%, respectively. In other words, the efficiency takes a significant hit if the switches are not sized for near-full charge transfer completion.

7.3.3 Extrinsic Losses

Previous chapters established parasitic coupling-, transistor driving-, and transistor leakage losses as three additional types of losses that are important to the design and optimization of switched-capacitor converters. In this section, their relevance in the presented topology are discussed.

The continuous charging of the parasitic coupling was shown to be a limiting factor to the achievable efficiency of regular SC converters in the monolithic context. From Fig. 7.7b, though, it can be appreciated that in the introduced topology, both top- and bottom-plate nodes are inherently soft-charged, which means that any parasitic coupling on these nodes is soft-charged as well. Moreover, because the parasitic coupling goes through nearly the same motions as the flying capacitor, their relative loss contribution is similar to the size of the coupling with respect to the flying capacitance. Considering that this relative coupling is typically between 1 and 7%, these losses are not as important for this converter type.

The transistor related losses, on the other hand, remain important loss contributors that provide a significant trade-off for the total transistor area. As was the case with the regular SC converter, these losses are considered extrinsic to the topology itself and are thus presumed not to interfere with its operation.

7.3.4 Optimization

The performance of the presented topology is decided by four major design parameters: the switch frequency, the total switch conductance per unit area, G'_{tot}, and the number of intermediate phases at the top-and bottom-side.

Because the output charge per cycle of the topology is nearly constant for larger M/N, the desired power-density basically sets the required switch frequency in stone:

$$f_{sw} \propto P_D. \tag{7.14}$$

This is a very different situation compared with the regular SC converter where the frequency is a key parameter to trade-off charge-sharing losses with transistor driving- or parasitic coupling losses. In other words, frequency is not an important optimization parameter, unless M/N are roughly speaking smaller than 10, which leaves G'_{tot} and M/N to optimize the converter.

To explore the optimal scaling of these parameters and the converter performance versus power-density, a closer look is taken at how the different loss contributors scale themselves. From the gyrator model, it can be shown that the intrinsic energy loss of the converter per cycle can be written as

$$E_{int} = \frac{C_{tot}}{M+1}\left(V_{in}^2 - 2V_{out}\frac{M+2}{M+1}(V_{in} - V_{out})\right), \tag{7.15}$$

assuming that $M = N$. Thus, E_{int} is approximately inversely proportionate to M for large enough M. This formula does not include the effect of incomplete settling due to the switch resistance. That being said, due to the significant efficiency degradation that was witnessed in Sect. 7.3.2 once the charge transfer completion percentage drops, it is presumed that the optimal switch sizing is large enough such that its effect on the intrinsic energy loss can be ignored. Furthermore, it is also assumed that the optimal charge transfer completion percentage stays approximately constant, which means that $RCf_{sw}p$, with p the number of phases and RC the charge transfer time constant, is constant as well. This leads to the following (approximated) relation:

$$\frac{M^2}{G'_{tot}} \propto \frac{1}{f_{sw}}. \tag{7.16}$$

Combining (7.14), (7.15) and (7.16) then yields

$$E_{int} \propto \sqrt{\frac{P_D}{G'_{tot}}}. \tag{7.17}$$

Because of (7.14), it can be appreciated that E_{int} has the same scaling as the intrinsic losses of the converter normalized by the output power, P_{int}. For the other loss contributors, the scaling is more easily identified. The transistor driving losses, P_{trans}, are proportionate to G'_{tot}, while the leakage energy loss, P_{leak}, is proportionate to G'_{tot}/f_{sw} and in turn to G'_{tot}/P_D. Consequently, it can be appreciated that at low power-densities, the leakage losses dominate the driving losses, and vice versa at high power-densities. Similarly to the regular SC converters, two distinct regimes of operation consequently emerge. For low P_D, optimization of the normalized total losses, P_N, yields

$$\begin{cases} G'_{tot} & \propto P_D \\ M & \propto 1 \\ P_N & \propto 1. \end{cases} \qquad (7.18)$$

In other words, an efficiency ceiling is reached where the intrinsic- and leakage losses balance each other out, and the number of soft-charging phases stays constant. At high power-densities, on the other hand, this behavior changes.

$$\begin{cases} G'_{tot} & \propto P_D^{1/3} \\ M & \propto P_D^{-1/3} \\ P_N & \propto P_D^{1/3}. \end{cases} \qquad (7.19)$$

Thus, both the losses and switch sizes scale with $P_D^{1/3}$, while M is inversely proportionate to the same factor. Note that this loss scaling is identical to that of regular switched-capacitor converters at high power-densities, as discussed in Chap. 2. Figure 7.12a shows the scaling of the optimized design parameters and energy loss, but using a full numerical optimization of the converter. As can be seen, both the high- and low power-density regimes are clearly demonstrated, and the simplified model used above remains valid for most of the power-density range. Ultimately, though, a point is reached where M gets close to 1 which causes some of the approximations that were used to break down. Here, a third regime can be witnessed where the optimal value for M is four.

The relative loss contribution of the same optimization is illustrated in Fig. 7.12b. It can be appreciated that the optimal relative contribution of P_{int} is approximately 65% regardless of the power-density, until the very high power-density regime is reached. The other 35% are attributed to a combination of leakage- and transistor driving losses, with the latter dominating at higher power-densities, as predicted.

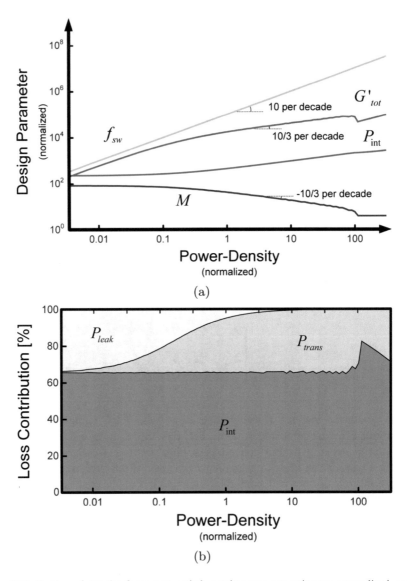

Fig. 7.12 Scaling of (**a**) the frequency, switch conductance per unit area, normalized power loss and number of soft-charging phases, and (**b**) the relative loss contribution of an optimized continuously-scalable-conversion-ratio SC converter for a VCR of 2:1

7.4   Implementation

The presented topology is implemented with $N, M = 32$, as illustrated in Fig. 7.13. To enable direct charge transfers between capacitors through the soft-charging nodes, the converter is split into $1 + N + M = 65$ phase-shifted cores. Thanks

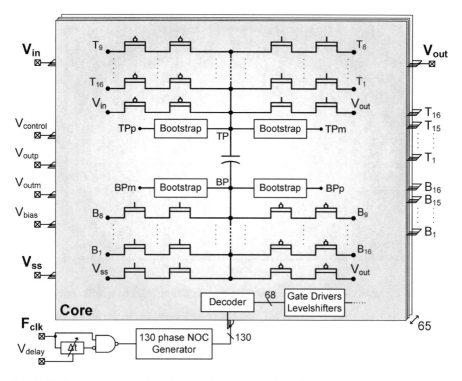

Fig. 7.13 System overview of implementation of proposed topology

to outphasing, only 16 top- and bottom nodes have to be implemented. Figure 7.14 demonstrates the behavior of these nodes versus time with outphasing. All switches are implemented using two stacked thin-oxide transistors to enable a wide V_{out} range with a V_{in} of 2 V. Furthermore, to maximize their conductance over the entire range, four 3 pF bootstrap capacitors are used per core, each of which generates a voltage rail relative to the top-/bottom plate, TP_p/TP_m and BP_p/BP_m, that is charged each time this plate connects to V_{out} using the external V_{outp} and V_{outm} rails. These external voltage rails are at a fixed voltage of 1 and -0.85 V relative to V_{out}. All transistors that connect to a soft-charging node need to block both positive and negative voltages across their drain–source and are therefore driven by the tristate buffers used in [MPK$^+$15].

An external frequency reference, F_{CLK}, is divided using the non-overlapping clock generator of Sect. 4.4.2 into 130 phases, which are then distributed to each core's local decoder. As such, the converter's frequency is $F_{CLK}/130$. The V_{ss}-connecting transistors together with the logic are powered by $V_{control} = 1.1$ V. $V_{control}$, V_{outp}, and V_{outm} are provided externally for flexibility and measurement purposes, but their peak current consumptions of 70, 8, and $-6\,\mu$A, respectively, allow them to be generated with classical charge-pumps and limited overhead.

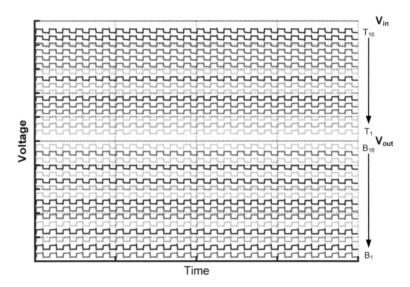

Fig. 7.14 Simulated waveforms of intermediate top- and bottom-side nodes, with outphasing

Table 7.2 Transistor width sizes per core in μm

	PMOS	NMOS
To V_{in}, V_{out} or V_{ss}	45.2	45.2
To intermediate nodes	13.56	13.56

7.5 Experimental Verification

To verify the working principle of the proposed topology, the design is realized in a 28 nm CMOS process using 9.5 nF of MOM- and thick-oxide MOS-capacitance for the flying capacitors, in addition to 780 pF for the bootstrap circuits. Similar to what is done in Chap. 6, the MOS capacitors are biased with a high-impedant bias, V_{bias}, of 3 V. The used transistor widths are given in Table 7.2. Figure 7.15 shows a micrograph of this realization, measuring 1.05 mm^2 in total. All cores are distributed over two rows to minimize the length of the top- and bottom-side intermediate nodes, as well as the control signals, both of which are positioned centrally.

7.5.1 Measurement Setup

Figure 7.16 illustrates the setup of the measurements performed in this section. Both the input- and output voltages are set using the Keithley 2450 source meters [Kei11] that simultaneously measure the corresponding currents. The external rails are similarly set using two-channel Keithley 2612A source meters [Kei17]. Critical to the operation of the implemented converter is that V_{outp} and V_{outm} are set correctly

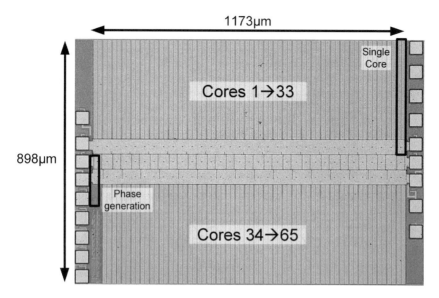

Fig. 7.15 Micrograph of the monolithic SC converter using a continuously-scalable-conversion-ratio topology

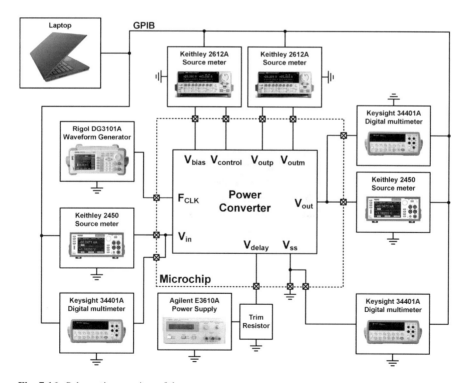

Fig. 7.16 Schematic overview of the measurement setup

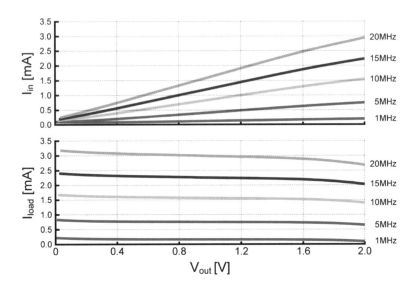

Fig. 7.17 Measured input- and output current versus output voltage for scaling F_{CLK}

relative to the output voltage. To this end, their voltages, together with all measurements, are controlled by a computer using a general purpose interface bus (GPIB).

7.5.2 Gyrator Behavior

First, the basic behavior of the converter is verified by measuring the converter's input- and load current versus the output voltage, with an input voltage of 2 V. The results hereof are shown in Fig. 7.17. Compared with Fig. 7.8, it can be appreciated that the measured results match the expected behavior reasonably well. The output current in particular stays close to 3mA for most of the output voltage range at the highest frequency, and only deviates at output voltages close to zero or the input voltage. In addition, both the input- and output current scale linearly with the applied reference clock frequency, which demonstrates the converter's compatibility with frequency control schemas.

7.5.3 Efficiency

Figure 7.18 plots the measured efficiency, including power consumed from all external rails, versus output voltage. The converter maintains a high efficiency over a wide VCR, staying above 80% efficiency from 0.43 to 2.13 V, and above 90% from 0.9 to 2.03 V. For output voltages larger than the input voltage of 2 V, the efficiency

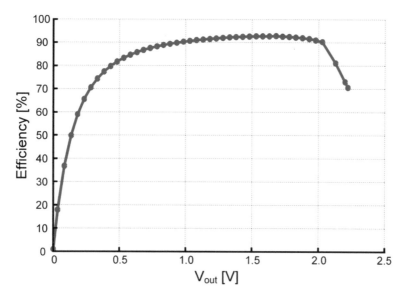

Fig. 7.18 Measured efficiency versus output voltage of the presented converter at $F_{CLK} = 20\,\text{MHz}$

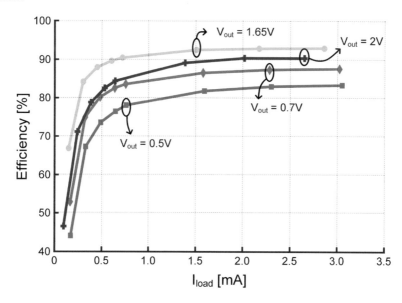

Fig. 7.19 Measured efficiency versus load current for several output voltages

drops rapidly. However, this is due to the fact that the converter was not designed to handle these output voltages, leading to body diode conduction.

The measured efficiency versus load current is shown in Fig. 7.19. Here, the load current is scaled by tuning the external clock frequency. A peak efficiency of 93%

is witnessed at a V_{out} of 1.65 V. Furthermore, the efficiency remains high down to 15–20% of the peak current, where the transistor leakage loss contribution becomes more significant.

7.5.4 Comparison

Table 7.3 compares the measurement results with the state of the art of monolithic SC converters. The presented work achieves the largest output voltage range of all compared designs, and does so using a single topology. In order to allow for fair comparison between different designs using different input- or output voltages, the concept of an efficient VCR range can be used. Here, the largest continuous VCR range for which the converter stays above a certain efficiency is considered. Also, as an additional normalization step, the input- and output voltages are swapped for step-up converters. From Fig. 7.20 it can be appreciated that the design has the highest VCR range above 80% efficiency, even when including gearbox converters that do not implement any kind of topology control.

7.6 Conclusions

In the first part of this chapter, it was pointed out that charge-sharing losses have forced all known switched-capacitor DC–DC converters until now to use topologies with fixed bias voltages, that consequently have inherent trade-offs between their voltage conversion ratio, efficiency, and output charge per cycle. In addition, it was shown that, at a fundamental level, soft-charging provided an opportunity to make efficient SC topologies with large voltage swing capacitors.

Next, a single-capacitor topology was derived using the concept of cornerstone phases. Said phases dictate the topology's basic structure, after which soft-charging phases can easily be implemented by adding intermediate nodes at the top-and bottom side. Furthermore, the topology was found to be compatible with outphasing, thereby reducing the switch losses and intermediate node area overhead.

The presented topology was modeled using a gyrator with three additional resistance elements, because, for constant input voltage, it delivers a near-constant output current regardless of the output voltage, even for step-up conversion. By increasing the number of intermediate soft-charging phases, the efficiency of the converter was found to tend to 100% regardless of the voltage conversion ratio. Parasitic coupling losses, which determine the maximum efficiency of regular SC converters, were additionally found to be insignificant in this topology.

A closer look was also taken at the topology's optimization leading to two distinct regimes, one at low- and one at high power-densities, being identified. While no closed-form analytic solution was arrived at, the optimal scaling of the important design parameters was derived. Similarly to regular converters, the

Table 7.3 Detailed comparison to state of the art

	This work [BS18b, BS19]	[BS16a]	[EDBC13]	[SM14b]	[JLMM18]	[JSB16]
Technology	28 nm	40 nm	130 nm	250 nm	65 nm	180 nm
Capacitors	MOM + MOS	MOM	Ferro electric	MIM	MOS + MIM	MIM
V_{in} [V]	2	1.855–2.07	1.5	2.5	0.22–2.4	2
V_{out} [V]	0–2.22	0.9	0.4–1.1	0.1–2.18	0.85–1.2	0.2–1.87
Number of topologies	1	1	4	15	24	79
VCR range[b] $\eta > 90\%$	0.56	0.04	0.05[a]	0	0	0.38
VCR range[b] $\eta > 80\%$	0.85	0.05	0.17[a]	0.24[a]	0.33[a]	0.7
Area [mm^2]	1.05	2.4	0.366	4.65	2.42	3.36
η_{peak}	93%	94.6%	93%	85%	84.1%	95%
Topology control?	Not needed	–	Yes	Yes	No	No
P_D at η_{peak} [mW/mm^2]	4.6	1.3	1.32	0.52[a]	13.2	0.13[a]

[a]Estimate based on graphs
[b]See definition in text

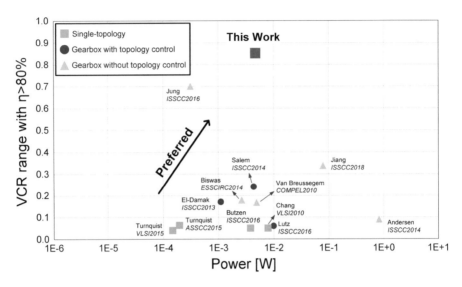

Fig. 7.20 Comparison with state of the art of fully integrated SC converters

proposed topology reaches a maximum efficiency at low power-densities, while at high power-densities the normalized losses scale with the cubic root of the power-density.

Finally, a circuit was fabricated in a 28 nm CMOS process that verified the working principle of the presented topology and advances the state of the art by achieving a peak efficiency of 93%, and a single-topology 0.9–2.03 V output voltage range with more than 90% efficiency.

Chapter 8
Conclusions

8.1 Overview of Work

The conversion of electrical energy from generator to consumer has become part of a foundation upon which much of today's society rests. For power converters used in electronics, there is a general drive to higher levels of integration, preferably onto the same microchip as the load. As explained in Chap. 1, this drive is partly fueled by a reduction of size and cost, both of which are key in enabling new applications such as the internet-of-things and micro-scale sensor nodes. Moreover, fully integrated DC–DC converters hold great promise in improving the system efficiency of higher-power microchips, such as mobile SoCs or server processor units, by reducing losses related to their power delivery network and enabling highly granular power domains with DVFS. Considering the fact that these applications are limited by their heat generation, any such improvement can be leveraged for better performance. However, to be a viable option, the integrated converter itself should have a sufficiently high efficiency, power-density, VCR, and a sufficiently large efficient VCR range.

Switched-capacitor converters are well-suited for the monolithic context due to the prevalence of both switches and capacitors in modern CMOS processes. Yet, the basic SC converter model of Chap. 2 demonstrated that their design space is constrained: at low power-densities a combination of parasitic coupling and low capacitance density limits their efficiency, while at high power-densities it is the balancing of transistor-related losses and capacitance density that causes an efficiency-power-density trade-off. In addition, charge-sharing losses inherently reduce the set of viable SC topologies to those with a narrow VCR range. In an attempt to enlarge the SC converter design space, this work proposed an extension on the commonly-used multiphasing technique, referred to as advanced multiphasing. Here, out-of-phase converter cores actively interact with each other and make use of the largely unexplored domain of phase/time.

© Springer Nature Switzerland AG 2020
N. Butzen, M. Steyaert, *Advanced Multiphasing Switched-Capacitor DC-DC Converters*, https://doi.org/10.1007/978-3-030-38735-8_8

Chapter 3 introduced a novel analysis method for DC–DC power converters, called voltage-domain analysis, where the time-averaged current distributions of converters and their elements are compared with each other in the voltage domain. By doing this analysis for an abstract SC converter, a general law of conventional SC DC–DC converters was derived that describes a fundamental relation between the flying capacitors' voltages, bottom-plate voltage swings, and charge multiplier elements. The voltage-domain analysis additionally provided insights into symmetries between certain two-phase SC topologies, and enabled a definition of topological efficiency. Surprisingly, this relation could also be used to show that multiphase SC converters have an intrinsic advantage over two-phase converters in terms of conversion efficiency at low power-densities.

A first advanced multiphasing technique, presented in Chap. 4 as Scalable Parasitic Charge Redistribution, focused on improving the efficiency of SC converters primarily at low power-densities. This was done by recycling charge on the parasitic coupling of the flying capacitors between several out-of-phase converter cores, using multiple charge redistribution steps. The basic model of Chap. 2 was expanded to include the effect of transistor leakage, which was in turn identified as another limiting factor to the efficiency at low power-densities, and important in determining the optimal efficiency with SPCR. Using this model, SPCR was demonstrated to improve the performance for a wide range of output powers. The working principles and effectiveness of SPCR were also verified with measurements. In fact, the prototype using SPCR still boasts the highest fully integrated DC–DC converter efficiency at time of writing at 94.6%. Furthermore, Chap. 5 explored SPCR as a means to generate multiple voltage rails for powering internal circuitry within a SC converter. The resulting parasitic multiple-input multiple-output converter was thoroughly analyzed and compared to similar SC converters, and had its behavior verified using measurement results.

The focus shifted towards improving the limited capacitance density with the introduction of two soft-charging techniques in Chap. 6, which has the inherent advantage of being useful both at high- and low-power-densities. Stage Outphasing achieved this by simply putting adjacent stages out of phase with each other, thereby not increasing the number of transistors for some topologies. But, the benefit of SO was shown to be inherently limited by the number of stages interacting with each other. Multiphase Soft-Charging, on the other hand, provides a scalable benefit, though at the cost of reducing the switch utilization. The Dickson converter was found to be an excellent candidate to use these techniques with, and in particular for higher VCRs. A realized 3:1 converter implemented both techniques to achieve a 60% higher effective capacitance density, and achieved 29 times larger output power-density compared with designs with similar efficiency in the literature.

Finally, Chap. 7 demonstrated that, while regular SC converters are required to have minimal capacitor voltage swing and thus fixed rational iVCRs due to charge-sharing losses, advanced multiphasing provides the opportunity to make topologies with large voltage swing capacitors that are made efficient through soft-charging. A particular topology was derived that has high efficiency regardless of the conversion ratio and behaves like a gyrator, making it the first capacitive DC–DC topology to achieve this. Its losses and optimization were carefully analyzed, revealing a low power-density regime where an efficiency ceiling is witnessed, and the emergence of an efficiency-power-density trade-off at higher power-densities similar to regular SC converters. The working principle of the topology was verified with a realization in a baseline CMOS technology that achieved the largest efficient VCR range to date.

Bibliography

[AKK+13] T.M. Andersen, F. Krismer, J.W. Kolar, T. Toifl, C. Menolfi, L. Kull, T. Morf, M. Kossel, M. Brändli, P. Buchmann, P.A. Francese, A 4.6W/mm^2 power density 86% efficiency on-chip switched capacitor DC-DC converter in 32 nm SOI CMOS, in *Applied Power Electronics Conference and Exposition (APEC), 2013 Twenty-Eighth Annual IEEE* (2013), pp. 692–699

[AKK+14] T.M. Andersen, F. Krismer, J.W. Kolar, T. Toifl, C. Menolfi, L. Kull, T. Morf, M. Kossel, M. Brändli, P. Buchmann, P.A. Francese, A sub-ns response on-chip switched-capacitor DC-DC voltage regulator delivering 3.7w/mm^2 at 90% efficiency using deep-trench capacitors in 32nm SOI CMOS, in *2014 IEEE International Solid-State Circuits Conference Digest of Technical Papers (ISSCC)* (2014), pp. 90–91

[AKK+17] T. Andersen, F. Krismer, J. Kolar, T. Toifl, C. Menolfi, L. Kull, T. Morf, M. Kossel, M. Brändli, P.A. Francese, A 10 W on-chip switched capacitor voltage regulator with feedforward regulation capability for granular microprocessor power delivery. IEEE Trans. Power Electron. **32**(1), 378–393 (2017)

[BWG+13] S. Bang, A. Wang, B. Giridhar, D. Blaauw, D. Sylvester, A fully integrated successive-approximation switched-capacitor DC-DC converter with 31mV output voltage resolution, in *2013 IEEE International Solid-State Circuits Conference Digest of Technical Papers* (2013), pp. 370–371

[BB48] J. Bardeen, W. Brattain, The transistor, a semi-conductor triode. Phys. Rev. **74**, 230–231 (1948)

[BS11a] Y. Beck, S. Singer, Capacitive transposed series-parallel topology with fine tuning capabilities. IEEE Trans. Circuits Syst. Regul. Pap. **58**(1), 51–61 (2011)

[BLC90] T. Berners-Lee, R. Cailliau, World Wide Web: Proposal for a HyperText Project (1990)

[BSW15] K.M. Bresniker, S. Singhal, R.S. Williams, Adapting to thrive in a new economy of memory abundance. Computer **48**(12), 44–53 (2015)

[BS09] T.V. Breussegem, M. Steyaert, A 82% efficiency 0.5% ripple 16-phase fully integrated capacitive voltage doubler, in *2009 Symposium on VLSI Circuits* (2009), pp. 198–199

[BS16a] N. Butzen, M. Steyaert, A 94.6%-efficiency fully integrated switched-capacitor DC-DC converter in baseline 40nm CMOS using scalable parasitic charge redistribution, in *2016 IEEE International Solid-State Circuits Conference (ISSCC)* (2016), pp. 220–221

© Springer Nature Switzerland AG 2020
N. Butzen, M. Steyaert, *Advanced Multiphasing Switched-Capacitor DC-DC Converters*, https://doi.org/10.1007/978-3-030-38735-8

[BS16b] N. Butzen, M. Steyaert, MIMO switched-capacitor converter using only parasitic capacitance with scalable parasitic charge redistribution, in *ESSCIRC Conference 2016: 42nd European Solid-State Circuits Conference* (2016), pp. 445–448

[BS16c] N. Butzen, M.S.J. Steyaert, Scalable parasitic charge redistribution: design of high-efficiency fully integrated switched-capacitor DC-DC converters. IEEE J. Solid State Circuits **51**(12), 2843–2853 (2016)

[BS17a] N. Butzen, M. Steyaert, A 1.1W/mm^2-power-density 82%-efficiency fully integrated 3:1 switched-capacitor DC-DC converter in baseline 28nm CMOS using stage outphasing and multiphase soft-charging, in *2017 IEEE International Solid-State Circuits Conference (ISSCC)* (2017), pp. 178–179

[BS17b] N. Butzen, M.S.J. Steyaert, Design of soft-charging switched-capacitor DC-DC converters using stage outphasing and multiphase soft-charging. IEEE J. Solid State Circuits **52**(12), 3132–3141 (2017)

[BS17c] N. Butzen, M.S.J. Steyaert, MIMO switched-capacitor DC-DC converters using only parasitic capacitances through scalable parasitic charge redistribution. IEEE J. Solid State Circuits **52**(7), 1814–1824 (2017)

[BS18a] N. Butzen, M. Steyaert, Proof of general switched-capacitor DC-DC converter law using voltage-domain analysis, in *2018 IEEE 19th Workshop on Control and Modeling for Power Electronics (COMPEL)* (2018), pp. 1–6

[BS18b] N. Buzen, M. Steyaert, A single-topology continuously-scalable-conversion-ratio fully integrated switched-capacitor DC-DC converter with 0-2.22V output and 93% peak-efficiency, in *2018 Symposium on VLSI Circuits* (2018), pp. 103–104

[BS19] N. Butzen, M. Steyaert, Design of single-topology continuously scalable-conversion-ratio switched- capacitor DC-DC converters. IEEE J. Solid State Circuits **54**(4), 1039–1047 (2019)

[BSS16] N. Butzen, A. Sarafianos, M. Steyaert, Monolithic switched-capacitor power converters: present trends and future predictions, in *International Workshop on Power Supply On Chip* (2016)

[Car14] F. Carobolante, Power supply on chip: from R&D to commercial products, in *International Workshop on Power Supply On Chip* (2014)

[CKP16] S.R. Challa, D. Kastha, A. Patra, A cascade point of load DC-DC converter with a novel phase shifted switched capacitor converter output stage. IEEE Trans. Power Electron. **31**(1), 353–368 (2016)

[CB95] A.P. Chandrakasan, R.W. Brodersen, Minimizing power consumption in digital CMOS circuits. Proc. IEEE **83**(4), 498–523 (1995)

[CFM$^+$10] L. Chang, D.J. Frank, R.K. Montoye, S.J. Koester, B.L. Ji, P.W. Coteus, R.H. Dennard, W. Haensch, Practical strategies for power-efficient computing technologies. Proc. IEEE **98**(2), 215–236 (2010)

[CR16] J. Cools, P. Reynaert, A 40nm bulk CMOS line driver for broadband communication, in *ESSCIRC Conference 2016: 42nd European Solid-State Circuits Conference* (2016), pp. 273–276

[Cop06] B.J. Copeland, *Colossus: The Secrets of Bletchley Park's Codebreaking Computers* (Oxford University Press, Oxford, 2006)

[DCVBDS12] N. De Clercq, T. Van Breussegem, W. Dehaene, M. Steyaert, Dual-output capacitive DC-DC converter with power distribution regulator in 90 nm CMOS, in *2012 Proceedings of the ESSCIRC (ESSCIRC)* (2012), pp. 169–172

[DLAH14] J. Delos, T. Lopez, E. Alarcón, M.A.M. Hendrix, On the modeling of switched capacitor converters with multiple outputs, in *2014 IEEE Applied Power Electronics Conference and Exposition - APEC 2014* (2014), pp. 2796–2803

[DGY$^+$74] R.H. Dennard, F.H. Gaensslen, H.-N. Yu, V.L. Rideout, E. Bassous, A.R. LeBlanc, Design of ion-implanted MOSFET's with very small physical dimensions. IEEE J. Solid State Circuits **9**, 256–268 (1974)

[Dic76]   J.F. Dickson, On-chip high-voltage generation in MNOS integrated circuits using an improved voltage multiplier technique. IEEE J. Solid State Circuits **11**(3), 374–378 (1976)

[Dum53]   G.W.A. Dummer, Electronic components in Great Britain. Electr. Eng. **72**(2), 167–169 (1953)

[EDBC13] D. El-Damak, S. Bandyopadhyay, A.P. Chandrakasan, A 93% efficiency reconfigurable switched-capacitor DC-DC converter using on-chip ferroelectric capacitors, in *2013 IEEE International Solid-State Circuits Conference Digest of Technical Papers (ISSCC)* (2013), pp. 374–375

[EBA+13] H. Esmaeilzadeh, E. Blem, R. St. Amant, K. Sankaralingam, D. Burger, Power challenges may end the multicore era. Commun. ACM **56**(2), 93–102 (2013)

[FKC+13] M. Fojtik, D. Kim, G. Chen, Y.S. Lin, D. Fick, J. Park, M. Seok, M.T. Chen, Z. Foo, D. Blaauw, D. Sylvester, A millimeter-scale energy-autonomous sensor system with stacked battery and solar cells. IEEE J. Solid State Circuits **48**(3), 801–813 (2013)

[Gol72]   H.H. Goldstine, *The Computer: from Pascal to von Neumann* (Princeton University Press, Princeton, 1972)

[GG46]   H.H. Goldstine, A. Goldstine, The electronic numerical integrator and computer (ENIAC). Math. Tables Other Aids to Comput. **2**(15), 97–110 (1946)

[Hoe60]   J.A. Hoerni, Planar silicon diodes and transistors, in *1960 International Electron Devices Meeting*, vol. 6 (1960), pp. 50–50

[Hol16]   W.M. Holt, Moore's law: a path going forward, in *2016 IEEE International Solid-State Circuits Conference (ISSCC)* (2016), pp. 8–13

[HL15]   Z. Hua, H. Lee, A reconfigurable dual-output switched-capacitor DC-DC regulator with sub-harmonic adaptive-on-time control for low-power applications. IEEE J. Solid State Circuits **50**(3), 724–736 (2015)

[ILPY17] Y. Ismail, H. Lee, S. Pamarti, C.K.K. Yang, A 36-V 49% efficient hybrid charge pump in nanometer-scale bulk CMOS technology. IEEE J. Solid State Circuits **52**(3), 781–798 (2017)

[ITR15]   ITRS. ITRS Reports (2015)

[JLH+15] J. Jiang, Y. Lu, C. Huang, W.H. Ki, P.K.T. Mok, A 2-/3-phase fully integrated switched-capacitor DC-DC converter in bulk CMOS for energy-efficient digital circuits with 14% efficiency improvement, in *2015 IEEE International Solid-State Circuits Conference – (ISSCC) Digest of Technical Papers* (2015), pp. 1–3

[JLMM18] Y. Jiang, M.K. Law, P.I. Mak, R.P. Martins, A 0.22-to-2.4V-input fine-grained fully integrated rational buck-boost SC DC-DC converter using algorithmic voltage-feed-in (AVFI) topology achieving 84.1% peak efficiency at 13.2mW/mm^2, in *2018 IEEE International Solid - State Circuits Conference - (ISSCC)* (2018), pp. 422–424

[JOB+14] W. Jung, S. Oh, S. Bang, Y. Lee, Z. Foo, G. Kim, Y. Zhang, D. Sylvester, D. Blaauw, An ultra-low power fully integrated energy harvester based on self-oscillating switched-capacitor voltage doubler. IEEE J. Solid State Circuits **49**(12), 2800–2811 (2014)

[JSB16]   W. Jung, D. Sylvester, D. Blaauw, A rational-conversion-ratio switched-capacitor DC-DC converter using negative-output feedback, in *2016 IEEE International Solid-State Circuits Conference (ISSCC)* (2016), pp. 218–219

[Kar15]   R. Karadi, Synthesis of switched-capacitor power converters: an iterative algorithm, in *2015 IEEE 16th Workshop on Control and Modeling for Power Electronics (COMPEL)* (2015)

[KP14]   R. Karadi, G.V. Pique, 3-phase 6/1 switched-capacitor DC-DC boost converter providing 16V at 7mA and 70.3% efficiency in 1.1mm^3, in *2014 IEEE International Solid-State Circuits Conference Digest of Technical Papers (ISSCC)* (2014), pp. 92–93

[Kei11]   Keithley Instruments, Inc., Series 2400 SourceMeter User's Manual (2011)

[Kei17]   Keithley Instruments, Inc., Series 2600B System SourceMeter Instrument Reference Manual (2017)

[Key07]   Keysight Technologies, Inc., Operating and Service Manuals for E361XA Bench DC Power Supplies (2007)

[Key12]   Keysight Technologies, Inc., 34401A Digital Multimeter User's Guide (2012)

[Key17]   Keysight Technologies, Inc., Keysight InfiniiVision 4000 X-Series Oscilloscopes User's Guide (2017)

[Key18]   Keysight Technologies, Inc., N2795A/N2796A Single-Ended Active Probes User's Guide (2018)

[Kil76]   J.S. Kilby, Invention of the integrated circuit. IEEE Trans. Electron Devices **23**(7), 648–654 (1976)

[KGWB08] W. Kim, M.S. Gupta, G.Y. Wei, D. Brooks, System level analysis of fast, per-core DVFS using on-chip switching regulators, in *IEEE 14th International Symposium on High Performance Computer Architecture, 2008. HPCA 2008* (2008), pp. 123–134

[KCB+15] N. Kurd, M. Chowdhury, E. Burton, T.P. Thomas, C. Mozak, B. Boswell, P. Mosalikanti, M. Neidengard, A. Deval, A. Khanna, N. Chowdhury, R. Rajwar, T.M. Wilson, R. Kumar, Haswell: a family of IA 22 nm processors. IEEE J. Solid State Circuits **50**(1), 49–58 (2015)

[LVJ17]   D.O. Larsen, M. Vinter, I. Jorgensen, Switched capacitor DC-DC converter with switch conductance modulation and pesudo-fixed frequency control, in *ESSCIRC 2017 - 43rd IEEE European Solid State Circuits Conference* (2017), pp. 283–286

[LSA11]   H.P. Le, S.R. Sanders, E. Alon, Design techniques for fully integrated switched-capacitor DC-DC converters. IEEE J. Solid State Circuits **46**(9), 2120–2131 (2011)

[LCSA13] H.P. Le, J. Crossley, S.R. Sanders, E. Alon, A sub-ns response fully integrated battery-connected switched-capacitor voltage regulator delivering 0.19W/mm^2 at 73% efficiency, in *2013 IEEE International Solid-State Circuits Conference Digest of Technical Papers* (2013), pp. 372–373

[LTZ+15] S.K. Lee, T. Tong, X. Zhang, D. Brooks, G.Y. Wei, A 16-core voltage-stacked system with an integrated switched-capacitor DC-DC converter, in *2015 Symposium on VLSI Circuits (VLSI Circuits)* (2015), pp. C318–C319

[LYM+17] Y.T. Lin, W.H. Yang, Y.S. Ma, Y.J. Lai, H.W. Chen, K.H. Chen, C.L. Wey, Y.H. Lin, J.R. Lin, T.Y. Tsai, Unsymmetrical parallel switched-capacitor (UP-SC) regulator with fast searching optimum ratio technique, in *ESSCIRC 2017 – 43rd IEEE European Solid State Circuits Conference* (2017), pp. 287–290

[MP17]   Y. Mahnashi, F.Z. Peng, Systematic approach to optimal SC converter synthesis for multi voltage-gain-ratio applications, in *2017 IEEE 18th Workshop on Control and Modeling for Power Electronics (COMPEL)* (2017), pp. 1–5

[Mak12]   M.S. Makowski, A note on resistive models of switched-capacitor DC-DC converters: unified incremental-graph-based formulas given, in *2012 International Conference on Signals and Electronic Systems (ICSES)* (2012), pp. 1–4

[MM95]   M.S. Makowski, D. Maksimovic, Performance limits of switched-capacitor DC-DC converters, in *26th Annual IEEE Power Electronics Specialists Conference, 1995. PESC '95 Record.*, vol. 2 (1995), pp. 1215–1221

[MD99]   D. Maksimovic, S. Dhar, Switched-capacitor DC-DC converters for low-power on-chip applications, in *30th Annual IEEE Power Electronics Specialists Conference, 1999. PESC 99*, vol. 1 (1999), pp. 54–59

[MBS13]   H. Meyvaert, T. Van Breussegem, M. Steyaert, A 1.65 W fully integrated 90 nm bulk CMOS capacitive DC-DC converter with intrinsic charge recycling. IEEE Trans. Power Electron. **28**(9), 4327–4334 (2013)

[MSBS14] H. Meyvaert, A. Sarafianos, N. Butzen, M. Steyaert, Monolithic switched-capacitor DC-DC towards high voltage conversion ratios, in *2014 IEEE 15th Workshop on Control and Modeling for Power Electronics (COMPEL)* (2014), pp. 1–5

[MPK⁺15] H. Meyvaert, G. V. Pique, R. Karadi, H.J. Bergveld, M.S.J. Steyaert, A light-load-efficient 11/1 switched-capacitor DC-DC converter with 94.7% efficiency while delivering 100 mW at 3.3 V. IEEE J. Solid State Circuits **50**(12), 2849–2860 (2015)

[Moo65] G. Moore, Cramming more components onto integrated circuits. Electron. Mag. **38**(8), 33–35 (1965)

[MW66] E. Moore, T. Wilson, Basic considerations for DC to DC conversion networks. IEEE Trans. Mag. **2**(3), 620–624 (1966)

[PC17] A. Paidimarri, A.P. Chandrakasan, A buck converter with 240pW quiescent power, 92% peak efficiency and a $2x10_6$ dynamic range, in *2017 IEEE International Solid-State Circuits Conference (ISSCC)* (2017), pp. 192–193

[PSS16] S. Pasternak, C. Schaef, J. Stauth, Equivalent resistance approach to optimization, analysis and comparison of hybrid/resonant switched-capacitor converters, in *2016 IEEE 17th Workshop on Control and Modeling for Power Electronics (COMPEL)* (2016), pp. 1–8

[Pew16] Pew Research Center, Smartphone Ownership and Internet Usage Continues to Climb in Emerging Economies (2016)

[Piq12] G.V. Piqué, A 41-phase switched-capacitor power converter with 3.8mV output ripple and 81% efficiency in baseline 90nm CMOS, in *2012 IEEE International Solid-State Circuits Conference* (2012), pp. 98–100

[RC10] Y.K. Ramadass, A.P. Chandrakasan, A batteryless thermoelectric energy-harvesting interface circuit with 35mv startup voltage, in *2010 IEEE International Solid-State Circuits Conference – (ISSCC)* (2010), pp. 486–487

[Ran06] A. Randall, Q&A: lost interview with ENIAC co-inventor J. Presper Eckert (2006)

[RIG08] RIGOL Technologies, Inc., DG3000 Series Function/Arbitrary Waveform Generator (2008)

[Roh17] Rohde & Schwarz GmbH & Co. KG, R&S SMBV100A Vector Signal Generator Operating Manual (2017)

[RMMM03] K. Roy, S. Mukhopadhyay, H. Mahmoodi-Meimand, Leakage current mechanisms and leakage reduction techniques in deep-submicrometer CMOS circuits. Proc. IEEE **91**(2), 305–327 (2003)

[Rup18] K. Rupp, 42 Years of Microprocessor Trend Data (2018)

[SM14a] L.G. Salem, P.P. Mercier, A recursive switched-capacitor DC-DC converter achieving 2^N-1 ratios with high efficiency over a wide output voltage range. IEEE Journal of Solid-State Circuits **49**(12), 2773–2787 (2014)

[SM14b] L.G. Salem, P.P. Mercier, An 85%-efficiency fully integrated 15-ratio recursive switched-capacitor DC-DC converter with 0.1-to-2.2V output voltage range, in *Solid-State Circuits Conference Digest of Technical Papers (ISSCC), 2014 IEEE International* (2014), pp. 88–89

[SM15] L.G. Salem, P.P. Mercier, A footprint-constrained efficiency roadmap for on-chip switched-capacitor dc-dc converters, in *2015 IEEE International Symposium on Circuits and Systems (ISCAS)* (2015), pp. 2321–2324

[SLM16] L.G. Salem, J.G. Louie, P.P. Mercier, Flying-domain DC-DC power conversion. IEEE J. Solid State Circuits **51**(12), 2830–2842 (2016)

[SS15a] A. Sarafianos, M. Steyaert, Fully integrated wide input voltage range capacitive DC-DC converters: the folding Dickson converter. IEEE J. Solid State Circuits **50**(7), 1560–1570 (2015)

[SPSS15]   A. Sarafianos, J. Pichler, C. Sandner, M. Steyaert, A folding dickson-based fully integrated wide input range capacitive DC-DC converter achieving Vout/2-resolution and 71% average efficiency, in *2015 IEEE Asian Solid-State Circuits Conference (A-SSCC)* (2015), pp. 1–4

[SS15b]   C. Schaef, J.T. Stauth, A 3-phase resonant switched capacitor converter delivering 7.7 W at 85% efficiency using 1.1 nH PCB trace inductors. IEEE J. Solid State Circuits **50**(12), 2861–2869 (2015)

[SDS17]   C. Schaef, E. Din, J.T. Stauth, A Hybrid switched-capacitor battery management IC with embedded diagnostics for series-stacked Li-Ion arrays. IEEE J. Solid State Circuits **52**(12), 3142–3154 (2017)

[SRS18]   C. Schaef, J.S. Rentmeister, J. Stauth, Multimode operation of resonant and hybrid switched-capacitor topologies. IEEE Trans. Power Electron. **33**, 10512–10523 (2018)

[See09]   M.D. Seeman, A Design Methodology for Switched-Capacitor DC-DC Converters. Ph.D. Thesis, EECS Department, University of California, Berkeley (2009)

[SPSD05]   B. Serneels, T. Piessens, M. Steyaert, W. Dehaene, A high-voltage output driver in a 2.5-V 0.25- μ m CMOS technology. IEEE J. Solid State Circuits **40**(3), 576–583 (2005)

[Sho76]   W. Shockley, The path to the conception of the junction transistor. IEEE Trans. Electron Devices **23**(7), 597–620 (1976)

[SSP+09]   D. Somasekhar, B. Srinivasan, G. Pandya, F. Hamzaoglu, M. Khellah, T. Karnik, K. Zhang, Multi-phase 1ghz voltage doubler charge-pump in 32nm logic process, in *2009 Symposium on VLSI Circuits* (2009), pp. 196–197

[SAT16]   T. Souvignet, B. Allard, S. Trochut, A fully integrated switched-capacitor regulator with frequency modulation control in 28-nm FDSOI. IEEE Trans. Power Electron. **31**(7), 4984–4994 (2016)

[SMCL11]   P. Stanley-Marbell, V.C. Cabezas, R.P. Luijten, Pinned to the walls - impact of packaging and application properties on the memory and power walls, in *2011 International Symposium on Low Power Electronics and Design (ISLPED)* (2011), pp. 51–56

[SVBM+11]   M. Steyaert, T. Van Breussegem, H. Meyvaert, P. Callemeyn, M. Wens, DC-DC converters: from discrete towards fully integrated CMOS, in *2011 Proceedings of the ESSCIRC (ESSCIRC)* (2011), pp. 42–49

[STM+15]   M. Steyaert, F. Tavernier, H. Meyvaert, A. Sarafianos, N. Butzen, When hardware is free, power is expensive! Is integrated power management the solution? in *European Solid-State Circuits Conference (ESSCIRC), ESSCIRC 2015 – 41st* (2015), pp. 26–34

[SBM+]   M. Steyaert, N. Butzen, H. Meyvaert, A. Sarafianos, P. Callemeyn, T. Van Breussegem, M. Wens, DCDC performance survey [Online]. Available: http://homes.esat.kuleuven.be/~steyaert/DCDC_Survey/DCDC_PS.html

[SRSK16]   C.R. Sullivan, B.A. Reese, A.L.F. Stein, P.A. Kyaw, On size and magnetics: why small efficient power inductors are rare, in *2016 International Symposium on 3D Power Electronics Integration and Manufacturing (3D-PEIM)* (2016), pp. 1–23

[TS16]   C.K. Teh, A. Suzuki, A 2-output step-up/step-down switched-capacitor DC-DC converter with 95.8% peak efficiency and 0.85-to-3.6V input voltage range, in *2016 IEEE International Solid-State Circuits Conference (ISSCC)* (2016), pp. 222–223

[UIOH91]   F. Ueno, T. Inoue, I. Oota, I. Harada, Emergency power supply for small computer systems, in *1991 IEEE International Symposium on Circuits and Systems*, vol. 2 (1991), pp. 1065–1068

[VBS09]   T. Van Breussegem, M. Steyaert, A 82% efficiency 0.5% ripple 16-phase fully integrated capacitive voltage doubler, in *2009 Symposium on VLSI Circuits* (2009), pp. 198–199

[VBS10a] T. Van Breussegem, M. Steyaert, A fully integrated gearbox capacitive DC/DC-converter in 90nm CMOS: optimization, control and measurements, in *2010 IEEE 12th Workshop on Control and Modeling for Power Electronics (COMPEL)* (2010), pp. 1–5

[VBS10b] T. Van Breussegem, M. Steyaert, Compact low swing gearbox-type integrated capacitive DC/DC converter. Electron. Lett. **46**(13), 892–894 (2010)

[BS11b] T.M. Van Breussegem, M.S.J. Steyaert, Monolithic capacitive DC-DC converter with single boundary-multiphase control and voltage domain stacking in 90 nm CMOS. IEEE J. Solid State Circuits **46**(7), 1715–1727 (2011)

[VPWRB11] G. Villar-Pique, L. Warmerdam, F. Roozeboom, H.J. Bergveld, Voltage conversion circuit (2011). EP Patent App. EP20,090,171,834

[WCS07] M. Wens, K. Cornelissens, M. Steyaert, A fully-integrated 0.18 um CMOS DC-DC step-up converter, using a bondwire spiral inductor, in *ESSCIRC 2007 – 33rd European Solid-State Circuits Conference* (2007), pp. 268–271

[Wol69] D.H. Wolaver, Fundamental Study of DC to DC Conversion Systems. Ph.D. Thesis, Massachusetts Institute of Technology (1969)

[Wol72] D. Wolaver, Basic constraints from graph theory for DC-to-DC conversion networks. IEEE Trans. Circuit Theory **19**(6), 640–648 (1972)

Index

© Springer Nature Switzerland AG 2020
N. Butzen, M. Steyaert, *Advanced Multiphasing Switched-Capacitor DC-DC
Converters*, https://doi.org/10.1007/978-3-030-38735-8